<u>Disclaimer</u>

Book Title: Characterization of the Range Performance of a 3D Imaging System (NIST TN 1695)

Book Author: Geraldine S. Cheok; Kamel S. Saidi; Marek Franaszek; James J. Filliben; Nicholas Scott;

Book Abstract: A 3D imaging instrument is a non-contact measurement instrument used to rapidly obtain 3D information about a scene or object. The use of and the applications for these instruments have grown tremendously in the last decade. The expanded use of 3D imaging instruments has revealed a significant lack of commonly accepted methods and standards to both characterize and report the performance of these systems, and to develop confidence limits for the data and their end products. Industry, and in particular the construction sector, needs open, consensus-based standards regarding the performance and use of 3D imaging instruments for construction applications. To support this need, the National Institute of Standards and Technology (NIST) is developing the measurement science (e.g., performance data) required to facilitate the development of standards for 3D imaging instruments. As a part of this effort, NIST conducted experiments to characterize the range performance of 3D imaging instruments. The objective of the experiments was to evaluate the effects of various factors on the range error of a 3D imaging instrument. The factors evaluated were: range, angle-of-incidence (AOI), reflectivity, azimuth angle, method of obtaining range measurement (measure single point on target or scanning target), and target type (planar vs. spherical). The results of these experiments are presented in this report.

Citation: NIST TN - 1695

Keywords: 3D imaging system; angle-of-incidence; laser scanning; performance evaluation; planar target; range error; reflectivity; test method

NIST TN 1695

Characterization of the Range Performance of a 3D Imaging System

Geraldine S. Cheok
Kamel S. Saidi
Marek Franaszek
James J. Filliben
Nicholas A. Scott

National Institute of
Standards and Technology
U.S. Department of Commerce

Characterization of the Range Performance of a 3D Imaging System

Geraldine S. Cheok, *Engineering Laboratory*

Kamel S. Saidi, *Engineering Laboratory*

Marek Franaszek, *Engineering Laboratory*

James J. Filliben, *Information Technology Laboratory*

Nicholas A. Scott, *currently with National Aeronautics and Space Administration (formerly with the Engineering Lab)*

May 2011

U. S. Department of Commerce
Gary Locke, Secretary

National Institute of Standards and Technology
Patrick D. Gallagher, Director

Abstract

A 3D (3 Dimensional) imaging instrument is a non-contact measurement instrument used to rapidly obtain 3D information about a scene or object. The use of and the applications for these instruments have grown tremendously in the last decade. The expanded use of 3D imaging instruments has revealed a lack of commonly accepted methods and standards to both characterize and report the performance of these systems, and to develop confidence limits for the data and their end products.

Industry, and in particular the construction sector, needs open, consensus-based standards regarding the performance and use of 3D imaging instruments for construction applications. To support this need, the National Institute of Standards and Technology (NIST) is developing the measurement science (e.g., performance data) required to facilitate the development of standards for 3D imaging instruments.

As a part of this effort, NIST conducted experiments to characterize the range performance of 3D imaging instruments. The objective of the experiments was to evaluate the effects of various factors on the range error of a 3D imaging instrument. The factors evaluated were: range, angle-of-incidence (AOI), reflectivity, azimuth angle, method of obtaining range measurement (measure single point on target or scanning target), and target type (planar vs. spherical). The results of these experiments are presented in this report.

Keyword: 3D imaging system; angle-of-incidence; laser scanning; performance evaluation; planar target; range error; reflectivity; test method.

Contents

List of Tables

List of Figures

1. Introduction

A 3D (3 Dimensional) imaging instrument is a non-contact measurement instrument used to produce a 3D representation of an object or a site [1]. A common example of such a representation is a point cloud as shown in Figure 1b. Figure 1 shows a typical application for 3D imaging instruments, where an existing structure is measured and a 3D model is generated for purposes such as checking as-is-conditions vs. design and for documentation.

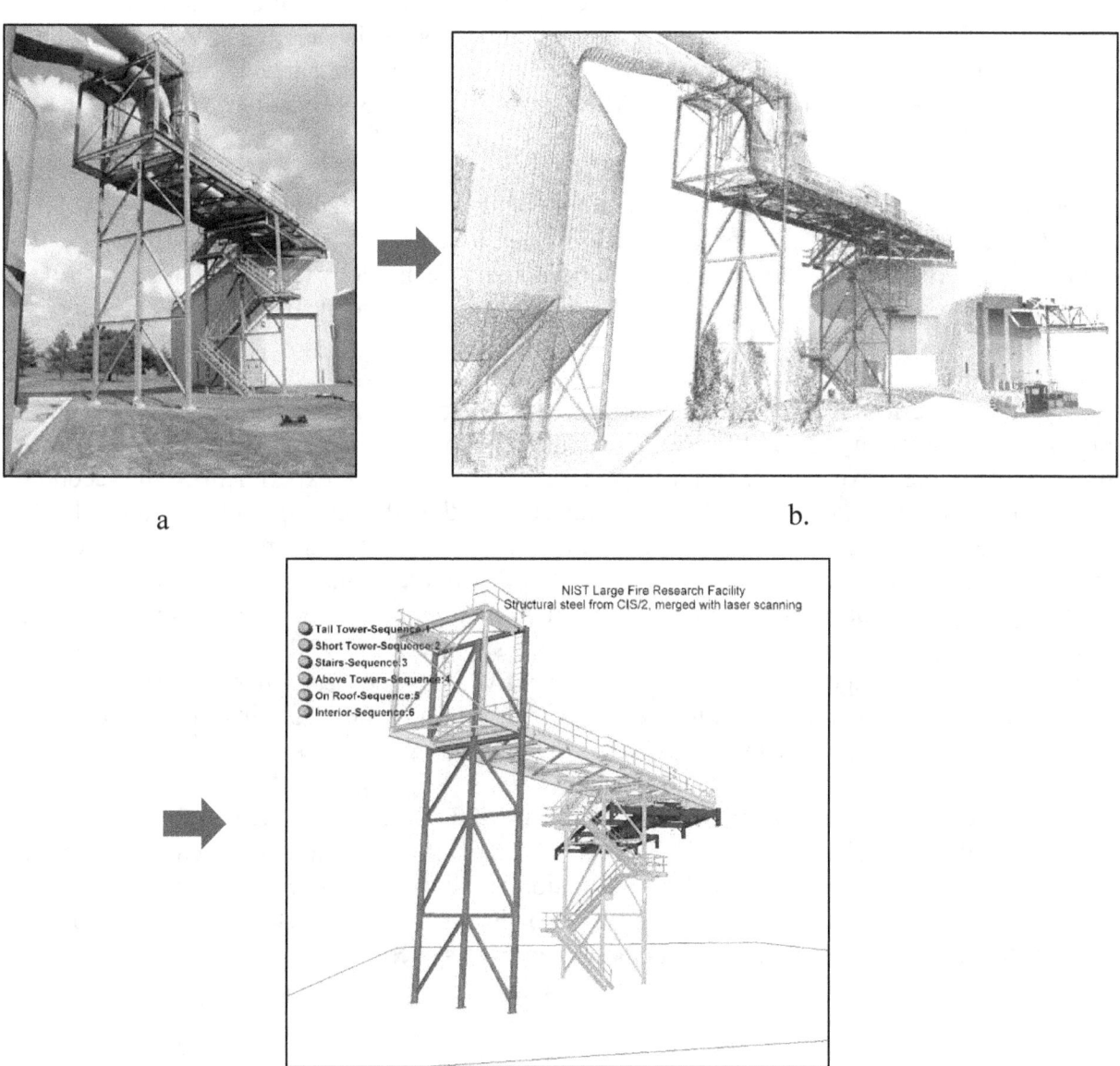

a

b.

c.

Figure 1. a. Digital photo of the structure. b. point cloud of the structure. c. VRML (Virtual Reality Mark Language) up representation of the structure.

1

The use of 3D imaging systems in widely varying fields such as manufacturing, forensics, autonomous vehicle navigation, and archeology grew significantly during the past decade. The greatest growth in the use of these systems was in the construction sector, where these new measurement technologies enable improved construction productivity through reduced errors and rework, schedule reduction, improved responsiveness to project changes, increased worker safety, and better quality control.

The expanded use of 3D imaging instruments has revealed a lack of commonly accepted methods and standards to both characterize and report the performance of these systems, and to develop confidence limits for the data and their end products. This lack of standards has been recognized ([2],[3], [4], [5]) and is being addressed by some standards development organizations. The main effort is the ASTM E57 Committee on 3D Imaging Systems [6]. Other efforts include VDI/VDE 2634 and ISO Technical Committee (TC) 172.

Industry, and in particular the construction sector, needs open, consensus-based standards regarding the performance and use of 3D imaging instruments for construction applications. To support this need, the National Institute of Standards and Technology (NIST) is developing the measurement science (e.g., performance data) required to facilitate the development of standards for 3D imaging instruments.

As a part of this effort, NIST has conducted experiments to characterize the range performance of 3D imaging instruments. The first experiments were conducted in 2002 [7]. These first sets of experiments were exploratory and helped guide the second set of experiments. The second set of experiments was conducted in 2008. The objective of the 2008 experiments was to evaluate the effects of various factors on the range error of a 3D imaging instrument. The factors evaluated were: range, angle-of-incidence (AOI), reflectivity, azimuth angle, method of obtaining range measurement (measure single point on target or scanning target), and target type (planar vs. spheres). The first three factors were investigated in 2002; the differences in the 2008 experiments were: longer range (160 m vs. 100 m), five known reflectivities (different colored paper targets of unknown reflectivities were used in 2002), and four AOIs instead of three. The last three factors were new factors investigated in the 2008 experiments.

Besides the efforts at NIST, other efforts to characterize the behavior of 3D imaging systems include [8], [9], [10], [5], [11], [12], [13], and [14]. In addition, manufacturers characterize their own instruments, but in most cases, this information is proprietary. The objective of the work presented in this report is to add to the existing knowledge on the characterization of 3D imaging systems to enable the development of standards for these systems.

2. Experiment Design and Test Procedures

The primary objective of these experiments was to determine the effect of range, angle-of-incidence, reflectivity, and azimuth on the range error of a 3D imaging instrument. The experiment design for these tests is presented in Section 2.1.1.

Secondary objectives of these experiments were to:

1. Determine if there are differences in the use of a planar target or a spherical target (Section 2.1.4),
2. Determine if there are differences in the measurement method, i.e., scanning the target or single point measurements (Section 2.1.3), and
3. Evaluate the test protocol in terms of practicality of tests, problems encountered, and lessons learned.

Both primary and secondary objectives support the efforts of ASTM E57.02 Test Methods Subcommittee in the development of a method for evaluating the range performance of a 3D imaging instrument.

The experiment design and variables are discussed in Section 2.1, and the test set up and test procedures are discussed in Section 2.2.

2.1 Experiment Descriptions

2.1.1 Experiment Design

The factors believed to have the greatest effect on the range error of a 3D imaging instrument were range, AOI, reflectivity, and azimuth angle. The different levels of the factors are shown below:

1. Range: 15 m, 60 m, 110 m, 160 m [(7.5, 30, 55, 80) % R_{max}[a]]
2. AOI: 0°, 20°, 40°, and 60°
3. Reflectivity: 20 %, 50 %, 75 % and 99 % (Planar targets, Section 2.2.1)
4. Azimuth angle: 1 = 1° (≤ 180°), 2 = 240°(> 180°)

For the first three factors, four levels were chosen because it was suspected, over the range of values for each factor, that the range error response would not be linear.

There were two reasons for having only two levels of the azimuth angle. If all four factors had four levels, the total number of tests based on all possible factor combinations would have been 256 (= 4 x 4 x 4 x 4). By reducing the number of azimuth levels from four to two, the total

---

[a] R_{max} = manufacturer specified maximum range for the 3D imaging system.

number of tests required is 128 for a full factorial design. It was felt that of the four factors, azimuth angle would have the least effect on range error.

In the ideal situation, the sequence for the 128 tests would have been randomized. However, this would have significantly increased the amount of time required to conduct the tests due to the time required to set up the target (e.g., moving the target and target holder after each test to a different range, removing and re-mounting target, re-alignment of target). Therefore, the 128 tests were grouped into 16 sets of eight tests, where for a given set, the target remained at a fixed distance (see Appendix A, Day 1 to Day 8).

Besides the 128 tests, additional tests were conducted. It is known that measurements of dark objects are problematic for many 3D imaging instruments. Therefore, 16 additional tests were planned where the target reflectivity was 2 %: four AOIs each at 15 m, 60 m, 110 m, and 160 m. Unfortunately, only one test at an AOI = 0° was performed at 15 m – tests at other AOIs were not conducted. All four AOIs were performed at 60 m. For the 110 m distance, extremely few measurements (approximately 10 points or less) were obtained from the target. Therefore, there was no data available from the 2 % reflective target at 110 m and 160 m.

Also, of the 128 tests, 16 tests were repeated. All the repeated experiments were conducted for azimuth angle = 240° (or level 2). For example, in Appendix A, Test 1r is a repeat of Test 103. The repeated tests were conducted to determine the reproducibility of the measurements.

2.1.2 Test Set A: Scanning a Planar Target

Some 3D imaging instruments allow for measurements of a single point on an object. Since this capability is not common for all instruments, a test protocol requiring measurements based on single point measurements would not have broad-scale value. As the basic method in which 3D imaging instruments acquire measurements involves scanning of an object, this method (scanning the target) was selected. However, the use of this method adds a level of complexity when post-processing the data. When measuring a single point, the determination of range to that point is trivial. However, when scanning a target, many points are acquired, and it is necessary to develop a method for defining the "correct" point on the target to determine the range to the target. The method used to define this point is described in Section 3. Once the point is defined, the range determination is straightforward.

The 128 tests, the 16 repeats and the tests using the 2 % reflective target, as described in Section 2.1.1, were performed by scanning a planar target. These tests constitute Test Set A. These tests constitute the majority of the tests in this experiment and are used to achieve the primary objective of this experiment.

2.1.3 Test Set B: Single Point Measurement Using a Planar Target

The 3D imaging system used for these experiments allows the user to point to a particular point and to obtain a measurement to that point (i.e., measure a single point on an object without the need to scan). This capability allows for the comparison of measurements obtained by either scanning a target or measuring single points. The instrument also has the capability of turning the laser on and off to allow the user to see where the instrument is pointing and what point is being measured.

For the single point measurement experiments, the target was placed at four distances (15 m, 60 m, 110 m, 160 m). Only two target reflectivities were used: 20 % and 99 %. At each distance and each target reflectivity, the target was rotated through the four AOIs (0°, 20°, 40°, 60°). These combinations totaled 32 experiments (see Appendix A, Single Point Measurements) and constitute Test Set B. The targets used in these tests were planar targets.

2.1.4   Test Set C: Measuring a Spherical Target

The use of spherical targets instead of planar targets is attractive as it eliminates the need to determine a point on the planar target as described in Section 2.1.2. When using spherical targets, a sphere is fitted to the acquired points and the range is then simply the distance from the instrument to the sphere center.

The spherical targets used in these experiments were custom-made anodized, aluminum, 152 mm diameter SMRs (spherically mounted retroreflectors). The use of these SMRs allowed the spheres to be measured with the 3D imaging instrument from one side, by scanning, and measured with a total station from the other side, by measuring to the retroreflector (see Figure 2). A sphere was placed at six distances: 15 m, 33 m, 60 m, 110 m, 142 m, and 160 m (see Appendix A, Spherical Target).

These measurements were used to determine if the range error was dependent on the type of target used (planar vs. spherical).

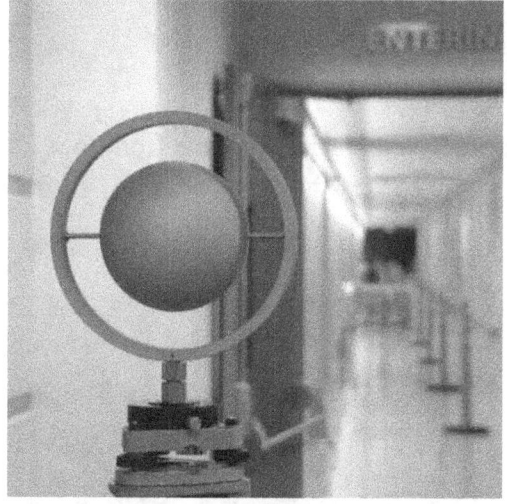 

Figure 2. 152 mm diameter SMR. Left image shows side facing the 3D imaging instrument. Right image shows side facing the total station and retroreflector in the center of the sphere.

2.2   Test Details, Set up, and Procedures

2.2.1   Planar Targets

The targets used for Test Sets A and B described in Section 2.1.1 and 2.1.3 were planar targets made of Spectralon[b] as shown in Figure 3. These targets were chosen because they are diffuse targets and the reflectivities are known. The reflectivities are also relatively uniform for wavelengths from 250 nm to 2500 nm (the wavelength of the instrument used in the experiments in this report is within this range). The dimensions of the target are 610 mm x 610 mm.

Figure 3.   A Spectralon target being mounted on a stand.

The targets were held in place by a specially-designed holder that ensures that the front face of the planar target is coincident with the stand's vertical rotation axis (see Figure 4).

---

[b] Certain trade names and company products are mentioned in the text or identified in an illustration in order to adequately specify the experimental procedure and equipment used. In no case does such an identification imply recommendation or endorsement by the National Institute of Standards and Technology, nor does it imply that the products are necessarily the best available for the purpose.

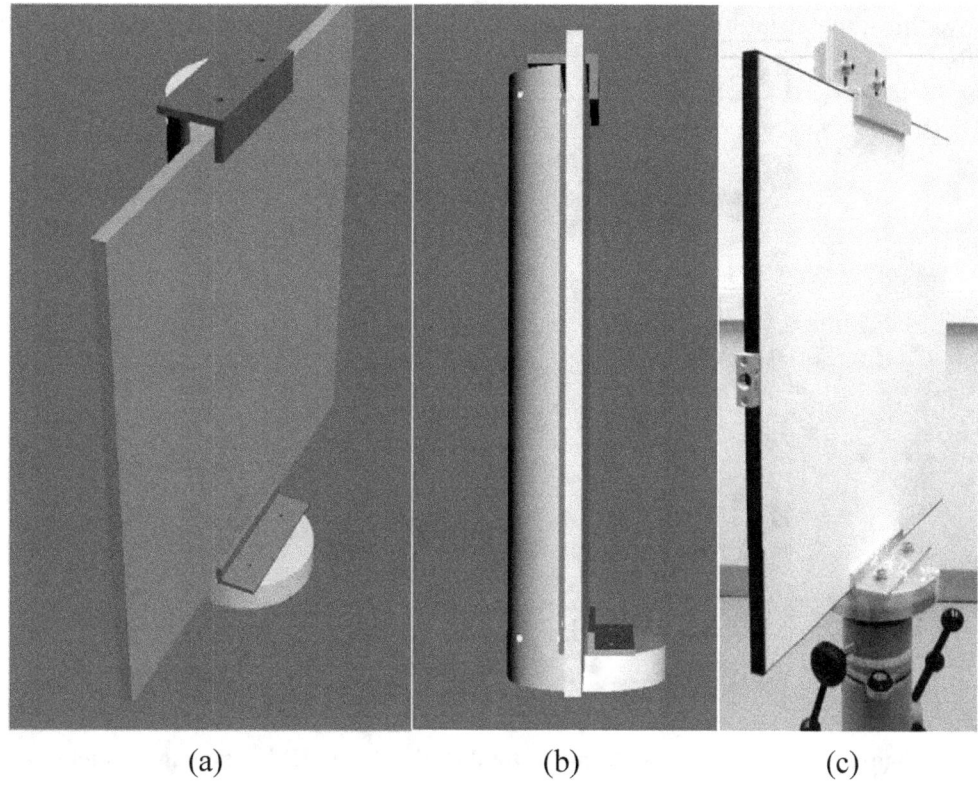

(a)   (b)   (c)

Figure 4.   The planar target in the target holder (a) CAD model oblique view; (b) CAD model side view; (c) photo.

Penetration of the laser into the Spectralon target causes bias in the measurement and is a known issue. Some experiments [15] were conducted prior to these range experiments to determine the amount of penetration into the Spectralon target. It was concluded in [15] that the amount of penetration into the Spectralon was within the noise of the 3D imaging instrument used for the experiments described in this report.

2.2.2   Test Location

The experiments were conducted in a hallway that connected two buildings where approximately 170 m clear distance was available (see Figure 5). This location afforded several other advantages. An important one was that it had a relatively low volume of pedestrian traffic, which meant less disruption to the building occupants and to the conduct of the experiments. Also, the hallways were relatively wide – again, reducing the disruption to the conduct of the experiments and to the building occupants.

Another advantage was that the two buildings connected by the hallway are part of the Advanced Metrology Laboratory at NIST. These buildings are about five stories underground (little vibration) and the temperature variation is small. The temperature and pressure were monitored at different times throughout the experiments. They were recorded at either the 3D imaging location or the target location. The average temperature over six days of tests was 19.2 °C and the standard deviation (s) was 0.4 °C. The average pressure was 100.38 kPa ($s = 0.6$ kPa). The maximum within-day variation was 1.7 °C and 0.4 kPa for the temperature and pressure, respectively.

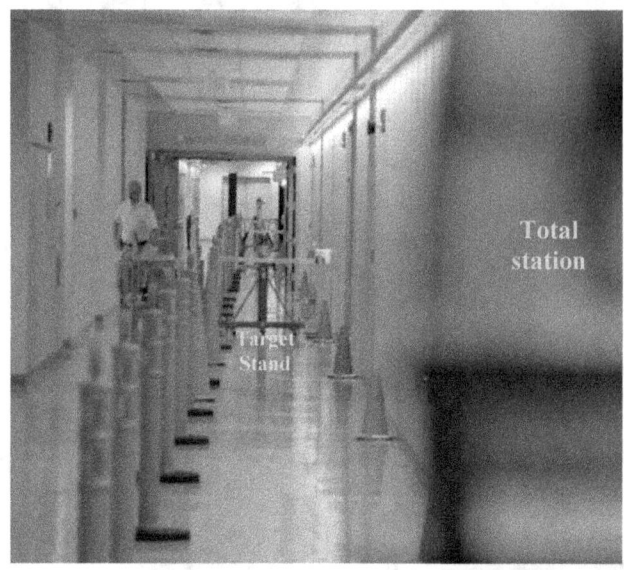

Figure 5. View of hallway from the total station to the 3D imaging instrument.

2.2.3   Ground Truth Measurements

The center of the 3D imaging instrument used in these experiments is "known"; that is, when the instrument is placed on a stand and centered over a point below the stand, the center of the instrument lies directly above that point (similar to other surveying instruments). This allows for the evaluation of absolute distance – distance from the 3D imaging system to the target. For 3D imaging instruments where the center of the instrument is unknown, the evaluation will have to be based on relative distance (distance between two targets).

To determine the absolute distance, three options for obtaining the ground truth distances were considered. These three options are shown in Figure 6.

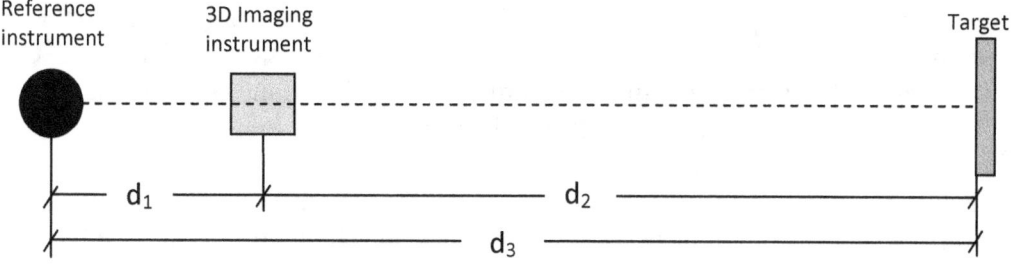

(a) Option 1: Ground truth distance = d_2 = $d_3 - d_1$

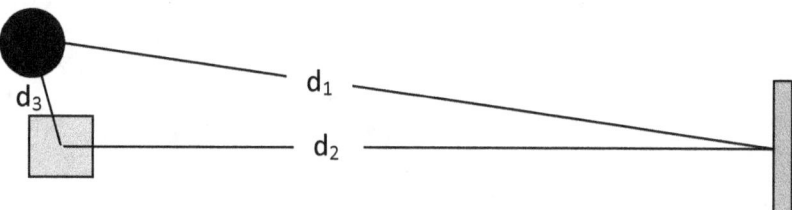

(b) Option 2: d_2 is obtained by registration of the reference instrument and 3D imaging instrument.

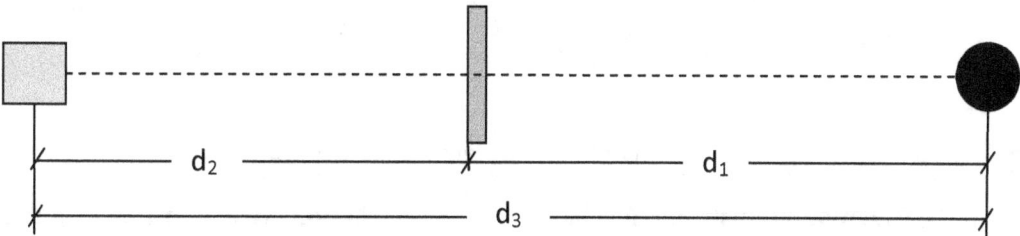

(c) Option 3: $d_2 = d_3 - d_1$ where d_1 is obtained by projecting a point from the back of the target to the front of the target.

Figure 6.   **Top view of three potential test set ups and the determination of ground truth measurements.**

Of the three options, Option 1 is the most direct method of determining the ground truth distance. The disadvantage of Option 1 is the need to dismount and then re-mount the 3D imaging instrument for each test as the 3D imaging system blocks the target from the reference instrument. This process would be a tremendous time expenditure as it involves re-leveling and realignment of the 3D imaging instrument each time. Also, the mounting and dismounting of the 3D imaging instrument adds to the variation in the data.

Option 2 involves registration of the coordinate frame of the reference instrument to the coordinate frame of the 3D imaging instrument. However, the registration process introduces errors and the magnitudes of these errors are unknown.

Option 3 involves measuring points on the back of the target and requires projecting the points onto the front of the target. The projection of these points introduces errors. However, Option 3 was selected because of the excessive time expenditure of Option 1 and because the error from the point projection was believed to be smaller than the registration error in Option 2.

Ground truth measurements of the distances between all the planar targets and the 3D imaging instrument were made using an industrial-grade total station. The total station used has a range uncertainty between ± 0.2 mm[c] and ±0.5 mm[d], and an angular uncertainty of ± 0.5° (as stated by the manufacturer). As described in the previous paragraph and as seen in Figure 7, the 3D imaging instrument was placed at one end of the hallway and the total station was placed at the opposite end.

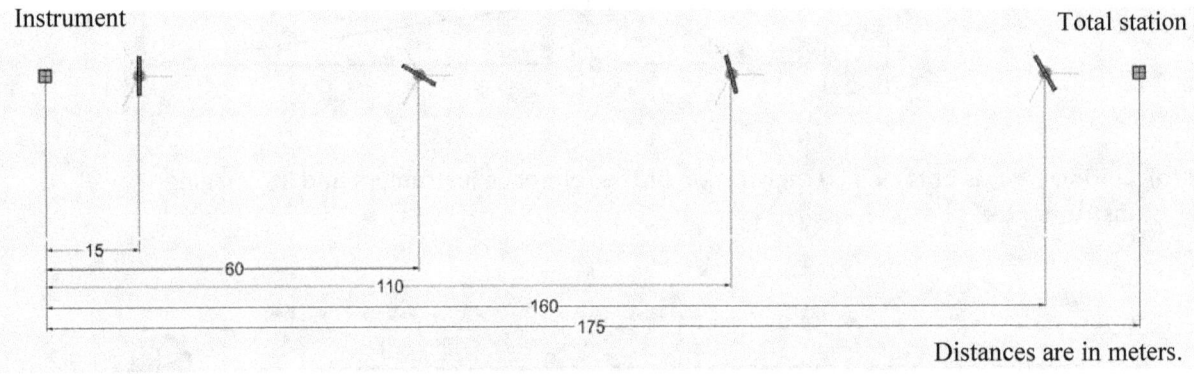

Figure 7. Top view of the layout of the total station, planar targets, and 3D imaging instrument.

The position of the 3D imaging instrument relative to the total station was determined each day by placing a 38.1 mm diameter spherically-mounted retro-reflector (SMR) on the 3D imaging system's stand and then measuring it with the total station.

The positions of the planar targets were determined by measuring (with the total station) the positions of four retro-reflective tape (RT) targets affixed to the backs of the planar targets. The four 25.4 mm square RT targets were placed in the corners of the backs of the planar targets. The centers of the RT targets were then measured (with a steel ruler) relative to the edges of the planar target (see Figure 8) so that they were equidistant from the geometric center of the planar target.

---

[c] When measuring a 38.1 mm spherically-mounted retro-reflector (SMR)
[d] When measuring a retro-reflective tape (RT) target.

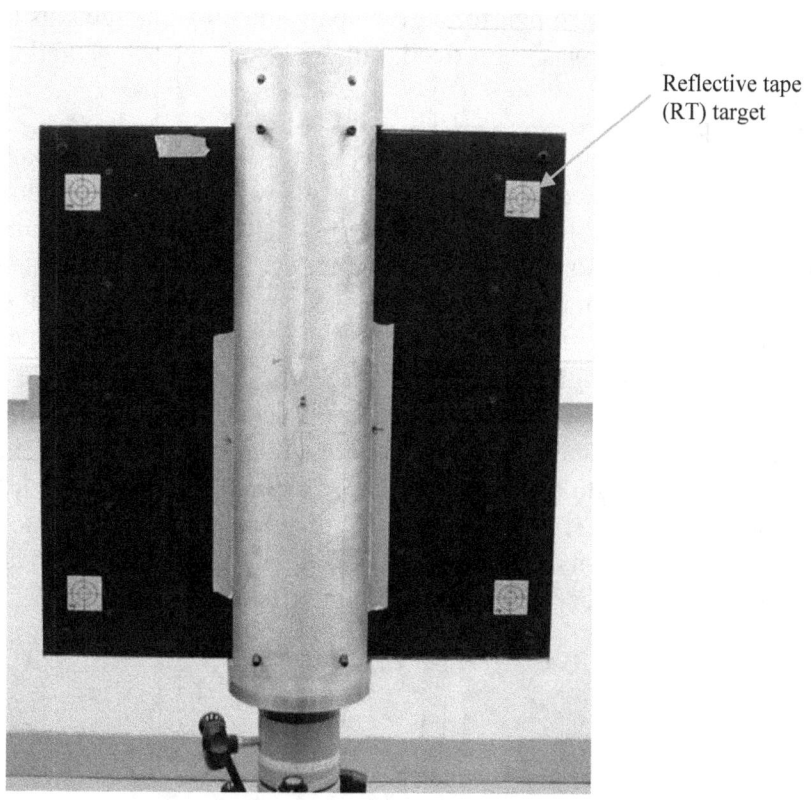

Reflective tape
(RT) target

Figure 8.   The reflective targets (RT) on the back of a planar target that is in the target holder.

The positions of the RT targets relative to the front surface of each of the planar targets as well as the thicknesses of each of the planar targets (Table 1) were determined using an instrument with a manufacturer-stated uncertainty of ± 100 μm. Measuring the positions of the four RT targets on the back of each planar target with the total station during the actual tests enabled the determination of the position of the center of the front surface of the planar target.

Table 1. Target Thickness

Target Reflectivity (%)	Thickness (mm)	s (mm)
2	17.3	0.19
20	16.9	0.05
50	17.0	0.06
75	17.0	0.06
99	17.5	0.08

For each test, when possible, the four RT targets were measured using both the front and back faces of the total station (10 times each) in order to eliminate any measurement bias due to the misalignment of the total station's rotation axes. It was not possible to measure the four targets when the AOI was 60° at the longer distances (110 m and 160 m). For these situations, the ground truth measurements for the next or previous test were used. This was possible because the tests were set up in pairs so that for a given pair, the only change was the AOI. This set up meant that the planar target was not moved, but only rotated on the stand about its vertical axis.

2.2.4   Test Set Up

The stands (Figure 7) for the instruments and targets were aligned by first aligning the two stands used to mount the total station and the 3D imaging instrument. Location marks (Figure 9) were placed under these two stands, and the stands were centered and leveled over these marks. The 3D imaging system was then placed over one stand and leveled. The total station was placed over the stand at the other end of the hallway (see Figure 5 and Figure 7) and also leveled. While the total station's vertical angle was fixed at zero (i.e., horizontal) and the azimuth angle was set so that the 3D imaging system was in the center of the total station's field-of-view, the height of the 3D imaging system's stand was adjusted until a horizontal reference line on the 3D imaging system (this marks the height of the instrument's optical center) became level with the total station sighting crosshairs. At this point, the 3D imaging system was removed from its stand and was replaced with a 38.1 mm diameter SMR. The total station's azimuth angle was then adjusted so that the SMR was at the center of the sighting crosshairs.

Figure 9.   Floor target with laser plummet on the center point.

The four target stands (Figure 10) were then aligned using the "line" from the total station to the 3D imaging system. Leveling (horizontal and vertical) of the stands and adjustments to the alignment line were an iterative process.[e]

---

[e] A problem was encountered when raising or lowering the target stands. Specifically, the top of the stand did not stay level. Potential reasons for this problem include the stands were not fabricated "plumb" and the bubble levels were incorrectly calibrated.

Figure 10.   Photo showing one of the metrology stands for targets and instruments.

2.2.5   Pre-Test Checks

At the beginning of each day, several checks were made. First, a check was made to ensure that the stand had not moved from the previous day. This was accomplished by making sure that both the 3D imaging instrument's stand and the ground truth instrument's stand were still centered over the location mark on the floor. Then, several targets, placed around the total station, were measured with the total station. The distances to these targets were used as another check for inadvertent movement of the stand from the previous day.

For the first day of experiments, the horizontal and vertical angles from the total station "zero" position to a marked point on the 3D imaging instrument were recorded. The vertical angle was used to set the height of the 3D imaging instrument and of the SMR that was used to measure the distance from the total station to the 3D imaging instrument.

2.2.6   Test Procedure

The settings for the 3D imaging system were: fixed focus at 50 m, point spacing of 5 mm x 5 mm at the target distance, and average of four measurements (i.e., each reported range measurement was the average of four measurements). These settings were used for all the experiments. The 5 mm x 5 mm setting resulted in about 14 000 to 15 000 points on the target.

At the beginning of each day of testing, the azimuth and elevation angles from the 3D imaging system to the total station were recorded. In addition, at the beginning or end of each set of tests (e.g., Day 1, Day 2, Repeats and 2 % reflectivity, Single Point Measurements), a 152 mm diameter SMR, placed at about 77 m from the 3D imaging system and slightly offset from the

13

target line, was scanned with the 3D imaging system and measured with the total station. This procedure was performed as a check of any inadvertent movement of the instrument stands.

For each of the test in Test Set A (Section 2.1.2), the planar target was scanned three times. Immediately after scanning the target, additional measurements were made only when time permitted. These measurements were single point measurements (five to ten) made to the front of the target. Note that these single point measurements are NOT the same as the single point measurements in Test Set B (Section 2.1.3).

For Test Set B, described in Section 2.1.3, 11 single point measurements were obtained for each of the 32 tests. When measuring the target, a visual check was made to ensure that the laser from the 3D imaging system was pointing at the approximate center of the target.

For Test Set C (spherical targets), the measurement procedure was straightforward. The SMRs were placed on the stands so that the spherical side faced the 3D imaging system and the retroreflector faced the total station – no additional alignment was made. The spherical face was scanned three times.

As the planar targets were measured (scanning or single point measurements), the ground truth measurements were made on the back of the targets (see Section 2.2.3) using a total station. For the spherical targets, the ground truth measurement was made by measuring the distance to the SMR using a total station.

The dates that the experiments were conducted are given in Table 2.

Table 2. Experiment Groupings and Test Dates.

Test Set	Experiment Grouping	Date of Test
A	Day 1	6/10/08
A	Day 2 and Day 3	6/11/08
A	Day 4 and Day 5	6/12/08
A	Day 6 and Day 7	6/13/08
A	Day 8, Repeats, 2 % reflective target	6/16/08
B	Single point measurements	6/17/08
C	Spherical Target	6/17/08

3. Data Post-Processing

The analyses of the data are based on determining the:

1. error in the range measurement where error = measurement – truth, and
2. RMS (root mean square) of the residuals of the geometric, plane or sphere, fit.

In these experiments, the range measurement is taken as the absolute distance or the distance from the 3D imaging system to a point on the target. The two main issues in the analyses are:

1. defining the "point on the target" and
2. whether the points used to determine the ground truth distance are the same as the points used to determine the target distance.

For these experiments, two methods were used to define the point on the target:

- Method 1A: the point is defined as the intersection of a line (see Section 3.3) and the target plane (see Section 3.1) and
- Method 1B: the point is defined as the geometric center of the target (see Section 3.1).

A schematic of the two different methods is shown in Figure 33 in Section 4.3. Method 1A was a candidate method that was discussed in the early development of a ranging protocol in the ASTM E57.02 Test Methods Subcommittee. A third method of defining the point on the target was the use of the centroid of points. However, this method was discarded because it was felt that the location of the centroid would be affected by the distribution of the points on the target – especially for a rotated target where there would be more points on the side of the target closer to the 3D imaging system.

The second issue arises from errors due to 1) misalignments of the targets, 3D imaging system, and the total station, and 2) the target not rotating about a vertical axis.

The data post-processing involves several steps. The first is the segmentation of the data (Section 3.1). This step involves automatically determining the points on the target, e.g., how outliers are defined. Once segmented, a plane is fitted through the points (Section 3.2). Another step is the determination of the measurement axis (Section 3.3) for Method 1A. This axis is used to intersect the target plane to obtain a point on the target. The third step is the determination of the ground truth measurements (Section 3.4).

3.1   Data Segmentation and Determining Geometric Center of Target

In the experiments, there were no other objects within 1 m of a target and its stand. When scanning a target, the scan region consisted of the whole target and a small region beyond the edges of the target. This resulted in some distant objects (e.g., wall) being included in the scans.

The first segmentation was performed to eliminate points from distant objects surrounding the target, e.g., wall behind target. For each dataset, a histogram of ranges was created where the size of a range bin was set to 20 % of the target width. The geometry of the scanned region ensured that the histograms had two main peaks well separated on a range axis: one peak corresponding to the points collected from the target and the other peak corresponding to the background behind the target. All other peaks contained less than 20 % of the data and were ignored. Of the two main peaks, the one with shorter ranges contained points acquired from the target – Subset 1 (see Figure 11a). Figure 11 gives a general flow and description of the data segmentation process used.

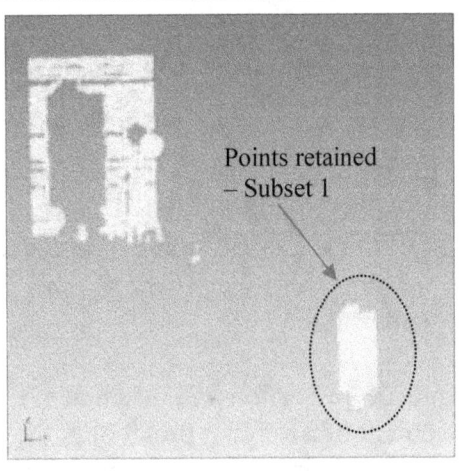

a. First segmentation to remove points in the background.

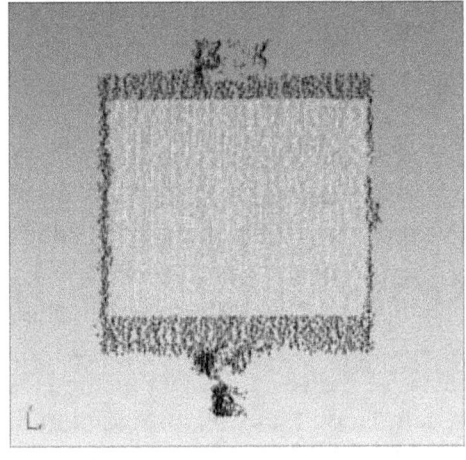

b. Second segmentation to obtain points, Subset 2 (green points), for initial plane fit.

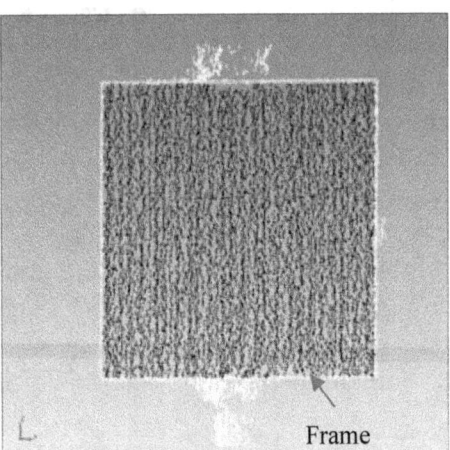

c. Third segmentation to determine the geometric center by using the points within the frame and excluding the points outside of the frame (white points).

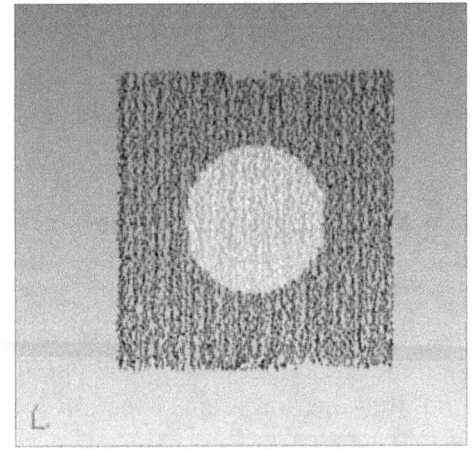

d. Fourth segmentation to obtain points, Subset 4 (yellow points), for second plane fit – target plane.

Figure 11. Sequence of segmentations to obtain set of points to create the target plane.

16

In addition to the points from the target, there were points from a small portion of the stand and from the target holder (see Figure 4). The second segmentation was performed to filter out these points. This was done by using the maximum Z and X coordinates and the minimum X coordinates of the data (see Figure 12). This information along with the height and the width of the target was used to define a region covering approximately 80 % of the flat target area (blue points in Figure 12) and was automatically selected. The set of points in this region is called Subset 2. An initial plane fitting (see Section 3.2) was performed using the points in Subset 2.

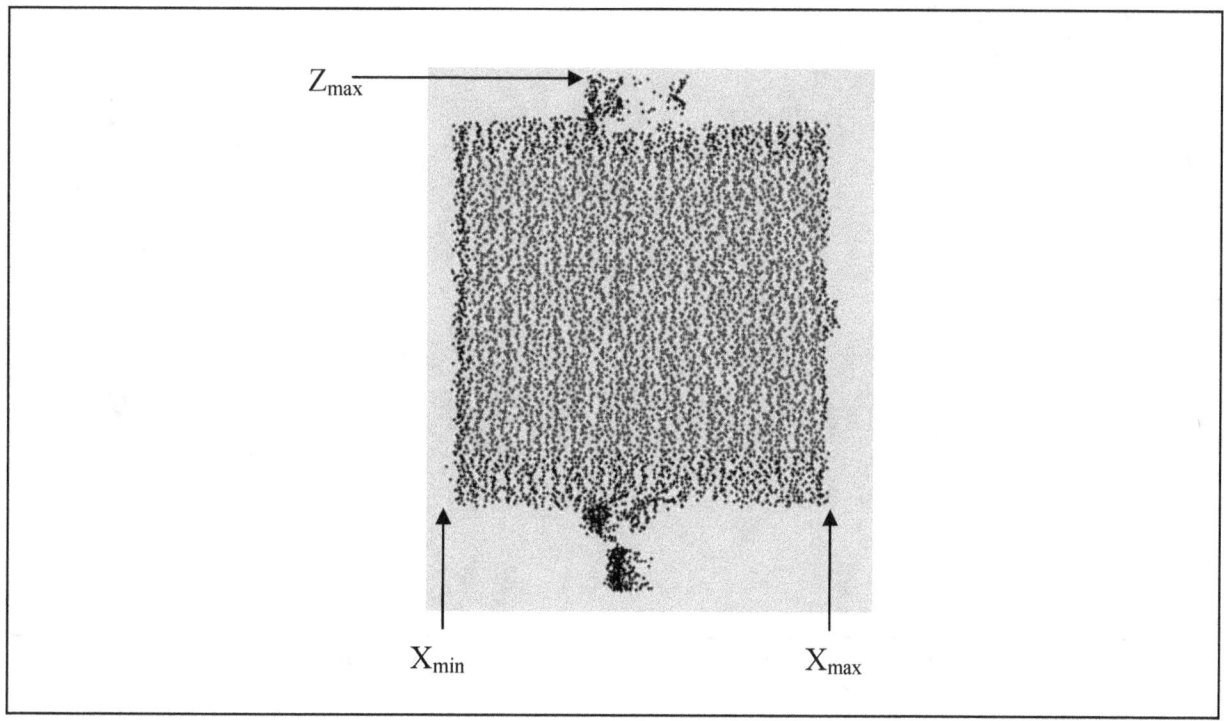

Figure 12. Second segmentation of data to obtain points from the target for the initial plane fit. Points in blue (Subset 2) were used for the initial plane fit.

A third segmentation of the data points was then conducted. The initial plane and knowledge of the target dimensions were used to determine the geometric center of the target. The basic procedure was to slide a frame in the plane defined by the initial plane fit and to count the number of points within the frame, see Figure 13. The following procedure was used:

1. Set the initial estimate of the geometric target center equal to the centroid of the points in Subset 2.
2. Using the dimensions of the target, construct a 2D frame based on the geometric center of the target. The corners of the frame had the following coordinates: $(X'_{min}, , Z'_{min})$, $(X'_{min}, , Z'_{max})$, (X'_{max}, Z'_{min}), and (X'_{max}, Z'_{max}).
3. Using all the points in Subset 1, determine and count the number of points within the frame. The points within this frame constitute the points in Subset 3.

4. Move the center of the frame by predefined increments of Δx and Δz.
5. Repeat steps 2 to 4 for a predefined number of times.

The derived geometric center of the target was set equal to the center of the frame for which the point count was the highest.

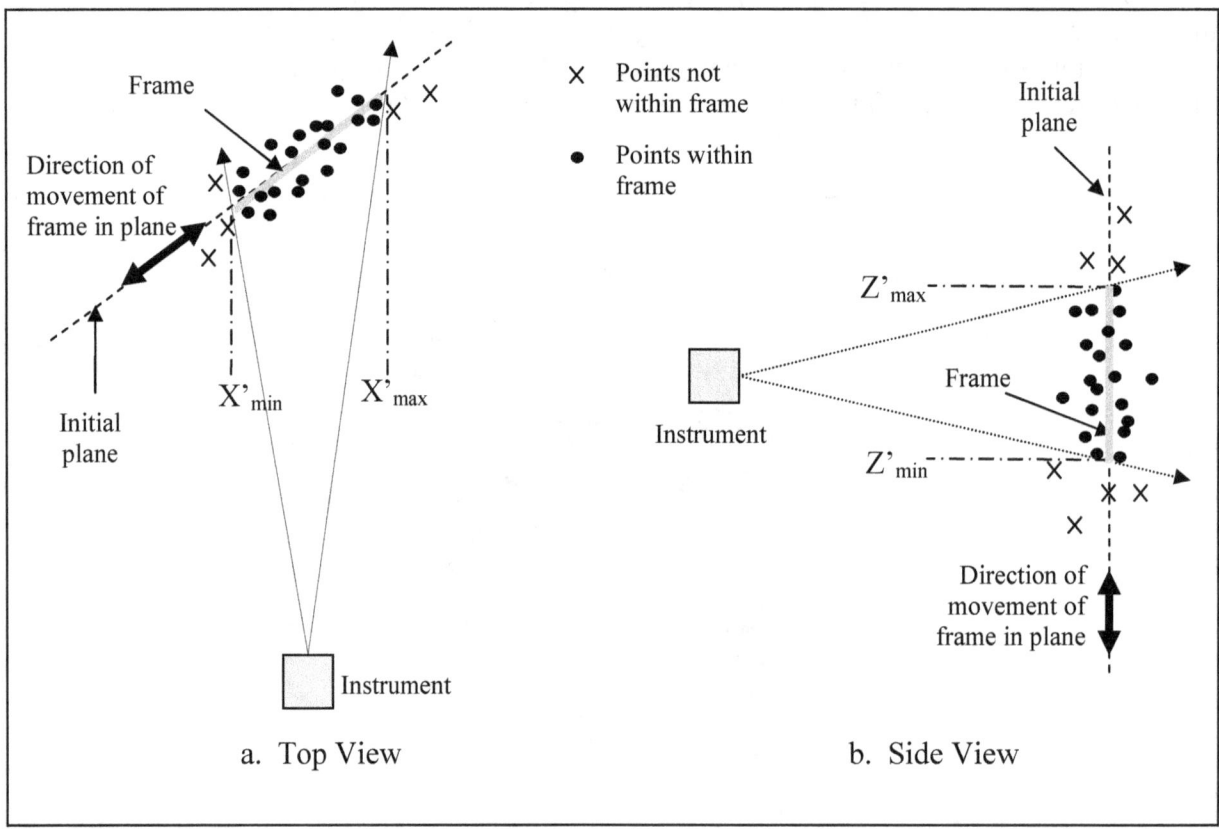

Figure 13. Second segmentation of the point cloud to locate the target.

To reduce edge effects, a fourth segmentation was performed to select points closer to the target center. To accomplish this, a cylinder centered on the geometric center, with a diameter equal to half the target width, and perpendicular to the initial plane was used for the fourth (final) segmentation of the data. Using the points in Subset 3, the points in Subset 4 consisted of the points within this cylinder (see Figure 11d). A second plane fitting was then performed using the points in Subset 4. This plane is considered the target plane.

3.2 Plane Fitting

The initial and second plane fitting described in Section 3.1 were performed as described in this section. Points P on plane satisfy the following equation:

$$w \bullet P(x, y, z) = D \tag{1}$$

where w is a unit vector perpendicular to the plane and D is the absolute value of the distance from the origin of coordinate system to the plane. Vector w can be defined by two angles: elevation, ϑ, and azimuth, φ (see Figure 14) as:

$$w(\vartheta, \varphi) = [\cos \vartheta \cos \varphi, \cos \vartheta \sin \varphi, \sin \vartheta]. \tag{2}$$

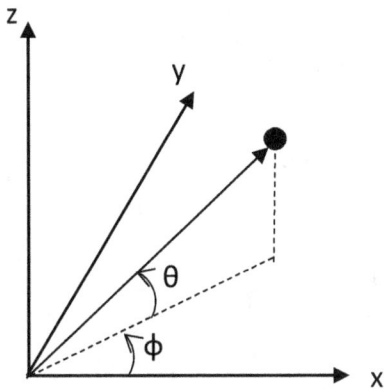

Instrument reference frame

Figure 14. Elevation and azimuth angles in instrument's reference frame.

Thus, a plane is defined by specifying three parameters: ϑ, φ, D. Having a set $P_{\{N\}}$ with N points $P_{\{N\}} = \{P_j, j = 1, \ldots, N\}$, a plane may be fitted to this dataset by minimizing the following error function Er,

$$Er(\vartheta, \varphi, D) = \frac{1}{N} \sum_{j=1}^{N} \left\| T_j - P_j \right\|^2, \tag{3}$$

where T_j is the intersection of a ray, through P_j, with the plane. Thus, T_j can be expressed as

$$T_j = t_j P_j \tag{4}$$

where t_j is a positive real number. Since T_j has to obey the plane equation (1), t_j can be calculated as

19

$$t_j = \frac{D}{r_j d_j} \tag{5}$$

where r_j is the j-th range and

$$\boldsymbol{P}_j(r_j, \vartheta_j, \varphi_j) = r_j \, \boldsymbol{p}_j(\vartheta_j, \varphi_j) \ , \quad \left\| \boldsymbol{p}_j \right\| = 1 \tag{6a}$$

$$d_j(\vartheta, \varphi) = \boldsymbol{w}(\vartheta, \varphi) \circ \boldsymbol{p}_j(\vartheta_j, \varphi_j) . \tag{6b}$$

In the above equations, ϑ_j and φ_j are the elevation and azimuth angles at which range r_j was acquired. Then, the error function in equation (3) can be expressed as

$$Er(\vartheta, \varphi, D) = \frac{1}{N} \sum_{j=1}^{N} \left(\frac{D}{d_j} - r_j \right)^2 . \tag{7}$$

When the error function is at its local minimum, the gradient of the function must be zero. In particular,

$$\frac{\partial Er(\vartheta, \varphi, D)}{\partial D} = 0 \tag{8}$$

and the parameter D may be explicitly calculated as a function of ϑ and φ

$$\boldsymbol{D}(\vartheta, \varphi) = \frac{\sum\limits_{j=1}^{N} r_j \, d_j^{-1}}{\sum\limits_{j=1}^{N} d_j^{-2}} \ . \tag{9}$$

By substituting the parameter D in equation (7) with the above function $D(\vartheta, \varphi)$, the error function is dependent only on ϑ and φ and can be expressed as:

$$Er(\vartheta, \varphi) = \frac{1}{N} \sum_{j=1}^{N} \left(\frac{D(\vartheta, \varphi)}{d_j(\vartheta, \varphi)} - r_j \right)^2 . \tag{10}$$

This error function was minimized using the variable metric method in Davidon-Fletcher-Powell (DFP) implementation, as described in [16].

3.3 Measurement Axis for Method 1A

There were several options to determine the measurement axis for Method 1A: 1) fit a line through the geometric center of the all the targets, 2) fit a line through the centroid of the target points (obtained as per Section 3.1) for all the targets, and 3) fit a line through the sphere centers (Section 2.1.4).

The first attempt was to define the measurement axis using the third option. The third option was chosen because it was felt that the uncertainties associated with the sphere centers was less than the uncertainties associated with the geometric centers of the targets. However, this resulted in errors on the order of meters in some instances (for combinations of higher AOIs and longer distances). These large errors indicate that the line through the spheres and the line through the targets were not parallel.

The second option was not selected because the centroid of the points on the target may be biased for a rotated target. There would be more points on the side closer to the instrument, and the centroid would be shifted towards the instrument.

The measurement axis was then defined using the first option – a line through the geometric centers of the targets. The geometric centers for each target were obtained as described in Section 3.1. For each day of tests, a line was fitted through all of these centers and the instrument origin (0,0,0). This line, the measurement axis, was used to intersect the target planes for those tests conducted on that day, i.e., there was a measurement axis created for each day of tests.

3.4 Ground Truth Measurements

The ground truth distance from the 3D imaging instrument to each target was defined as the difference between (1) the distance from the total station to the 3D imaging system and (2) the distance from the total station to the geometric center of the front of the planar targets, as shown in Figure 6c. The distance to the 3D imaging system was measured directly using an SMR as described in Section 2.2.4. The front of the target is the side of the target facing the 3D imaging system which is the Spectralon side of the target. Since the total station only measured the four RT targets on the back of the planar targets, the geometric center of the front of the planar targets was assumed to coincide with the projection of the centroid of the four RT targets along the normal of the back plane by a distance equivalent to the target thicknesses presented in Table 1. The projection of the geometric center from the back plane of the target to the front plane is shown in Figure 15.

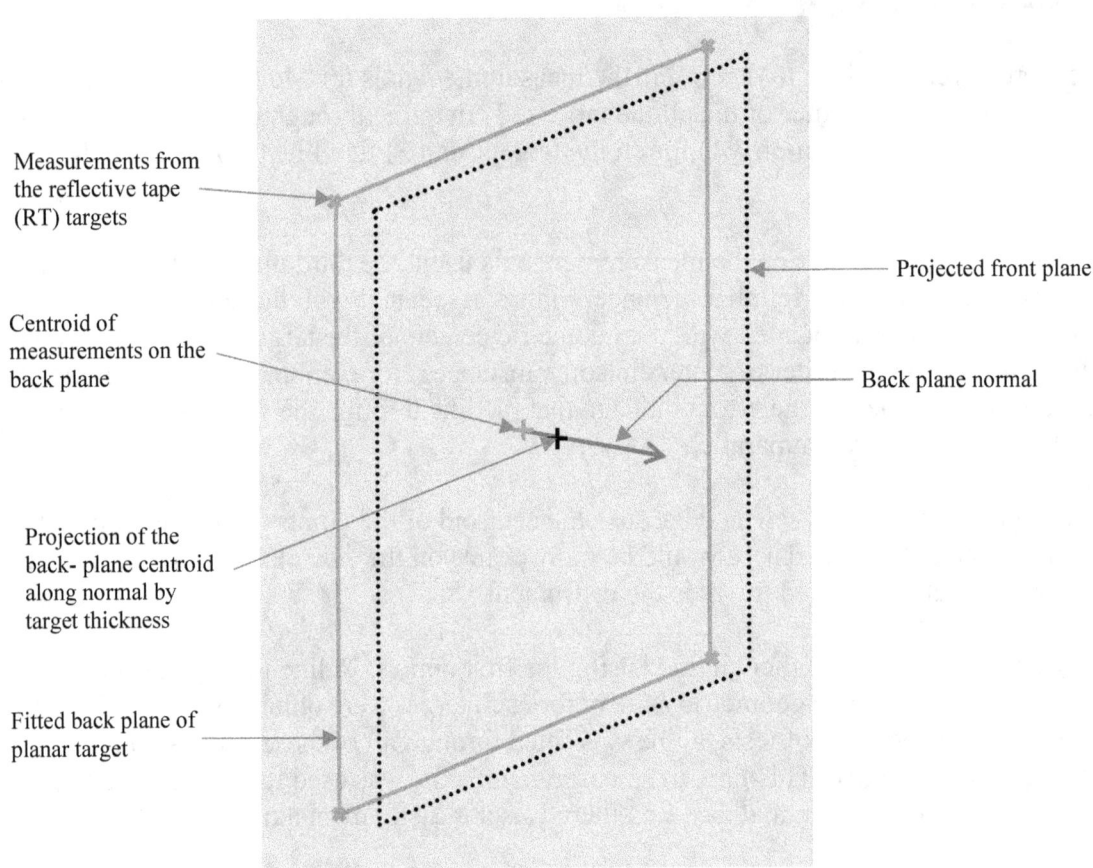

Figure 15. Calculation of the geometric center of the front of the planar target.

4. Data Analysis

4.1 Variability from Mounting/Dismounting 3D Imaging Instrument

As mentioned in Section 2.2.6, a 152 mm diameter SMR was slightly offset from the targets at about 77 m from the 3D imaging instrument and was measured at the beginning of each set of tests. In some cases, the SMR was scanned two or more times within a day - that is, the instrument was not moved between scans of the SMR. Comparisons between the derived sphere centers as obtained between days and within a day provide an indication of the magnitude of the error introduced by the removal of the 3D imaging instrument at the end of each day and re-installing it the next day.

The comparison made was based on the angle between the vectors from the instrument origin to the sphere center and instrument's Y-axis. The average of the angle between the vectors from two different days was 0.12° (n [= # of samples] = 7, s = standard deviation of data = 0.07°). The average angle between the vectors obtained within a single day was 0.05° (s = 0.06°, n = 4). Based on the t test and 95 % confidence limits, the two means are statistically equivalent. That is, any variation caused by moving the 3D imaging system on/off the stand is equivalent to the variation exhibited by the instrument under normal usage where the instrument was used to conduct scans for a period of time but was not removed from the stand.

4.2 Experimental Results from Test Set A

4.2.1 Main Factor Effects

This experiment examines the relative effect of the four main factors: range, AOI, reflectivity, and azimuth. It is of interest to determine the relative effect of each of the four individual factors on the range error and range noise. For a given factor, a "main factor effect" is a numeric value that represents how much on the average the error or noise changes from one level to the next.

For factors with two levels (azimuth), the computed effect is assigned to the factor as a whole. For factors with three or more levels (range, AOI, reflectivity), the factor effect may change (non-linearly) depending on the levels. If a factor effect changes depending on the setting of another factor, then interactions exist and they may be identified and quantified.

In any event, it is important to ascertain the nature and relative importance of the four factors. This will be done both graphically and quantitatively.

The range errors for the 128 tests, the 16 repeats, and the 2 % reflective target from Test Set A are shown in Figure 16 (Method 1A – point on target defined by the intersection of a line and a plane, see Section 3) and Figure 17 (Method 1B – point on target defined as the geometric center of the target, see Section 3). In the top row of plots in these figures, individual error values for each test are plotted. The total number of points (n) in each of the plots in the top row is 446.

In the bottom row of Figure 16 and Figure 17, the average error values are plotted. The factor effects are estimated by computing the differences of the averages for various levels within a factor. The average errors also give an indication of the instrument bias.

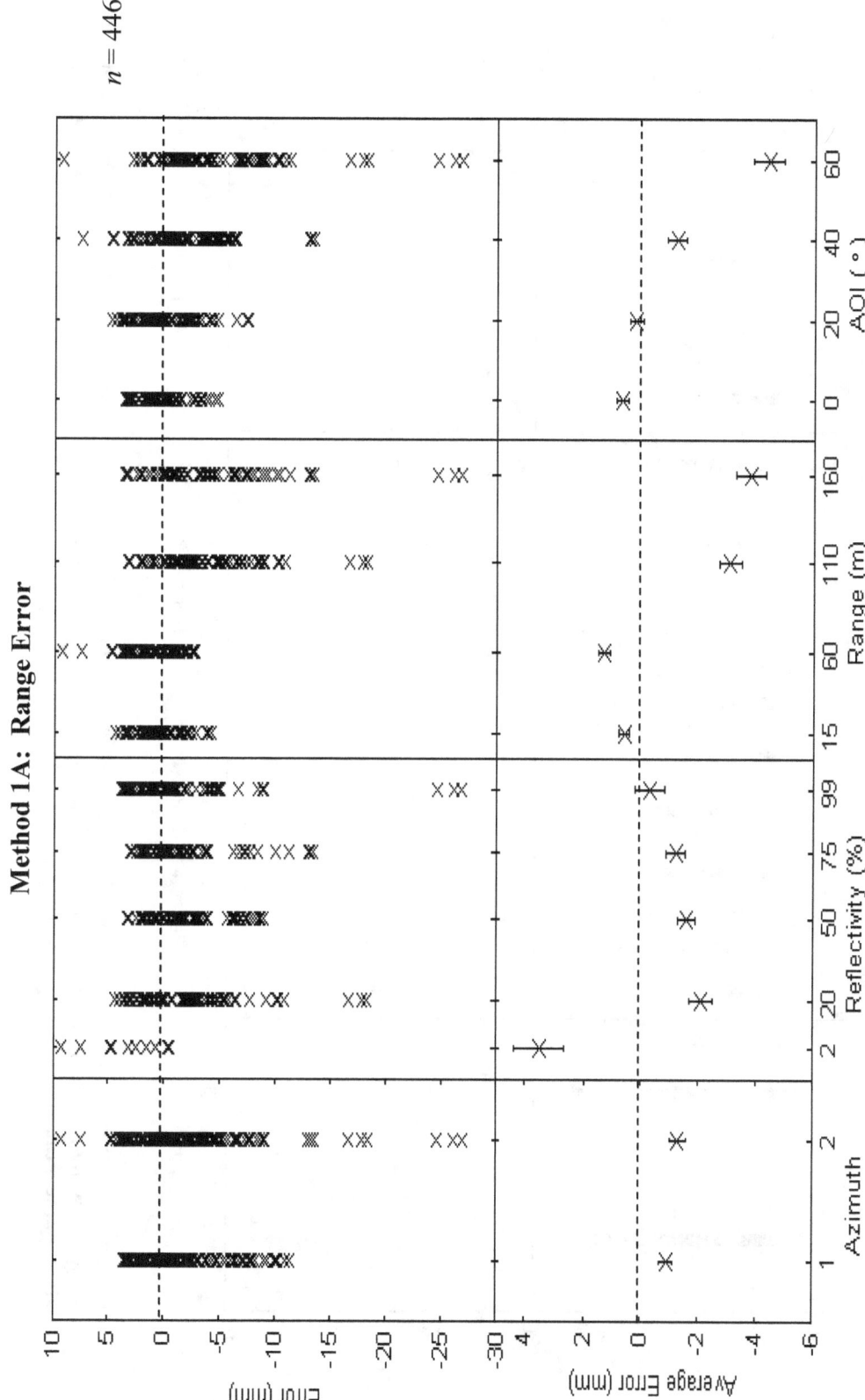

Figure 16. Method 1A (line plane intersection) - Main factor effects on range error. The top of plots are the error values versus each level of each factor. The bottom row of plots are the averages for each of the level of each factor. Note: Error bars in the bottom row of plots are the standard deviations of the mean.

25

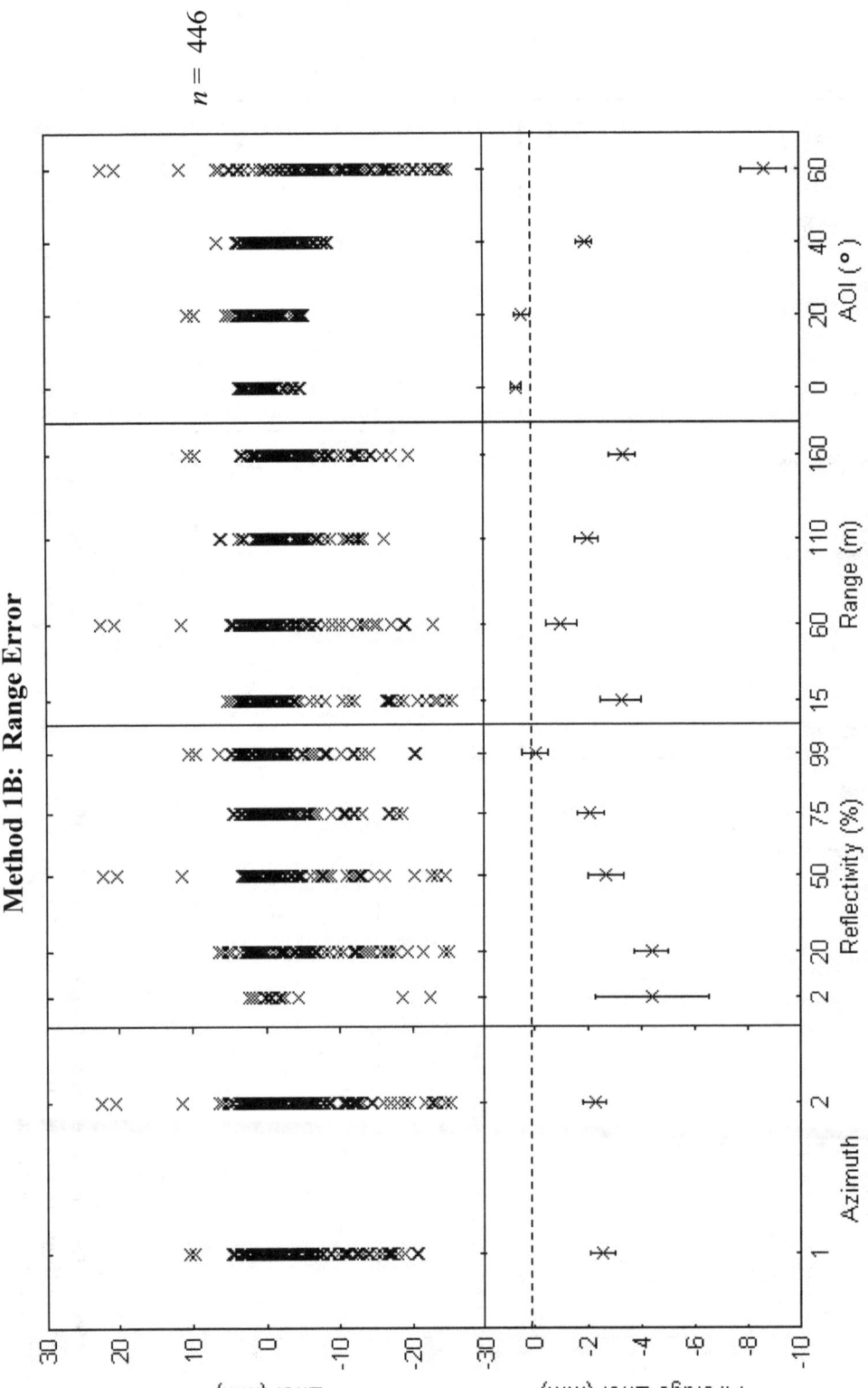

Figure 17. Method 1B (geometric center) - Main factor effects on range error. The top of plots are the error values versus each level of each factor. The bottom row of plots are the averages for each of the level of each factor. Note: Error bars in the bottom row of plots are the standard deviations of the mean.

In the top row of Figure 16 and Figure 17, larger errors are defined as |errors| ≥ 10 mm. From Figure 16 and Figure 17, the following observations are made:

Factor	Method 1A	Method 1B
Azimuth	the majority of the larger errors are associated with Azimuth 2	
	Azimuth 2 shows more variation in range error	Azimuth 2 shows more variation in range error
Range	all of the larger errors occur at 110 m (0.55 R_{max}) and 160 m (0.8 R_{max})	larger errors occur at all 4 ranges.
		there is a negative range bias, i.e., the instrument is underestimating the range
Reflectivity	the majority of the larger errors occur for reflectivities of 20 % and 99 %	
	except for the 2 % reflective target, the magnitude of the range error increases for decreasing reflectivity	the magnitude of the range error increases for decreasing reflectivity
	2 % reflective target shows the largest variation	2 % reflective target shows the largest variation
AOI	the majority of the larger errors are associated with an AOI of 60°	the majority of the larger errors are associated with an AOI of 60°
	significant increase in error at 60°	significant increase in error at 60°
Other observation	instrument bias = average range error of -1.1 mm (s = 4.2 mm)	instrument bias = average range error of -2.4 mm (s = 6.4 mm)

Based on the bottom row in Figure 16 and Figure 17, the following conclusions are made:

- Based on the difference (Δ) between the maximum and minimum averages in each plot, the ranking (from most to least effect on range error) of the factors are:

Method 1A	Method 1B
1) reflectivity (Δ = 5.6 mm)	1) AOI (Δ = 9.5 mm)
2 & 3) AOI and range (tied) (Δ = 5.1 mm)	2) reflectivity (Δ = 5.1 mm)
	3) range (Δ = 2.3 mm)
4) azimuth (Δ = 0.5 mm)	4) azimuth (Δ = 0.3 mm)

From the reflectivity plot, it is seen that the results for the 2 % reflective target significantly affected the ranking for Method 1A. If the 2 % reflective target were omitted, the ranking would be:

27

Method 1A (w/o 2 % results)	Method 1B (w/o 2 % results)
1) AOI (Δ = 5.4 mm)	1) AOI (Δ = 9.1 mm)
2) range (Δ = 4.8 mm)	2) reflectivity (Δ = 4.4 mm)
3) reflectivity (Δ = 1.8)	3) range (Δ = 2.8 mm)
4) azimuth (Δ = 0.8 mm)	4) azimuth (Δ = 0.5 mm)

- All of the average errors for both methods are within manufacturer's specified range uncertainty[f] of ± 7 mm.
- Azimuth: For both Methods 1A and 1B, the average range errors for Azimuths 1 and 2 are approximately equal indicating no or minimal effect of azimuth on range error.
- Range:
  - Method 1A - the instrument overestimates the range for shorter ranges [≤ 60 m (0.3 R_{max})] and underestimates the range for longer ranges [≥ 110 m (0.55 R_{max})].
  - Method 1B – the instrument underestimates the range over all ranges.
- Reflectivity:
  - Method 1A - except for the 2 % reflective target, the trend shows that the magnitude of the range errors increases as the reflectivity decreases, i.e., the underestimation of the range increases as the reflectivity decreases.
  - Method 1B – clear trend showing that the magnitude of the range errors increases as the reflectivity decreases
- AOI: For both Methods 1A and 1B, the trend shows a nonlinear increase of the magnitude of the range error with increasing AOI. There is little effect on range error for AOIs ≤ 20° and a significant increase for AOIs > 40°.

The RMS of the residuals of the plane fit (henceforth referred to as RMS for brevity) for the 128 tests, the 16 repeats, and the 2 % reflective target tests from Test Set A are shown in Figure 18. The total number of points (n) in each of the plots in the top row of the figure is 446. The RMS values provides an indication of the range noise.

---

[f] The term used in the manufacturer's specifications is accuracy. It is assumed that the specified accuracy is the uncertainty of the range measurement.

28

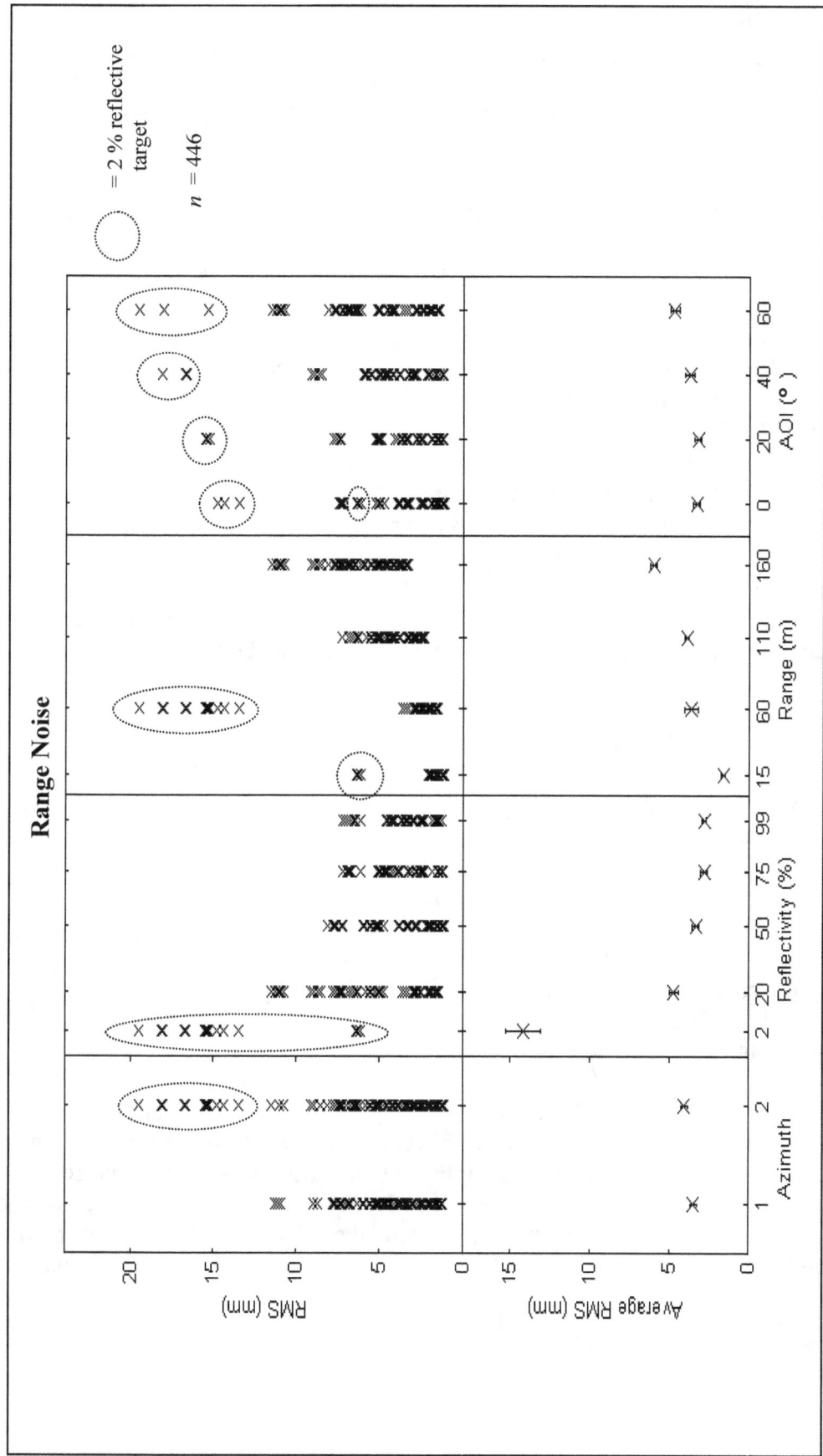

Figure 18. Main factor effects on range noise (RMS of the plane fit). The top of plots are the RMS values versus each level of each factor. The bottom row of plots are the averages for each of the level of each factor. Note: Error bars in the bottom row of plots are the standard deviations of the mean and some are not visible as they are smaller than the symbol.

29

The range errors from the 2 % reflective target, although among the larger range errors, did not stand out from the other range errors. In contrast, there is a separation (except for one test) between the RMS values for the 2 % reflective target and the values for the other reflectivities as indicated in Figure 18. Note that there were no data for the 2 % reflective target for ranges greater than 60 m.

Based on the top row of plots in Figure 18, the following conclusions are made:

- Azimuth: no effect on range noise (omitting the results from the 2 % reflective target since data for the 2 % reflective target were only obtained for Azimuth 2).
- Reflectivity: There is a nonlinear relationship of increasing range noise with decreasing reflectivity. The range noise is relatively unchanged for reflectivities > 50 % and begins to increase for reflectivities ≤ 50 %. There is a significant increase in instrument noise for a reflectivity of 2 %.
- Range: There is a linear relationship between range noise and range. The RMS values increase as range increases. The increases are significant for ranges ≥ 110 m.
- AOI: There is a nonlinear relationship between range noise and AOI. The range noise is relatively unchanged for AOI ≤ 20° and begins to increase for AOI ≥ 40°.

Based on the bottom row of plots in Figure 18 and on the differences (Δ) between the maximum and minimum values in each plot, the ranking, from most to least effect on range noise, of the factors are:

1) reflectivity (Δ = 11.3 mm)
2) range (Δ = 4.4 mm)
3) AOI (Δ = 1.5 mm)
4) azimuth (Δ = 0.6 mm)

If the 2 % reflectivity data are omitted, the ranking is:

1) range (Δ = 4.6 mm)
2) reflectivity (Δ = 1.9 mm)
3) AOI (Δ = 1.5 mm)
4) azimuth (Δ = 0.0 mm)

From Figure 16, Figure 17, and Figure 18, the factor having the least effect on range error and range noise is azimuth. The factor having the most effect on range error is reflectivity based on Method 1A and AOI based Method 1B. The factor having the most effect on range noise is reflectivity. As noted earlier, the results from the 2 % reflective target tests skewed the rankings. If the results from 2 % tests were omitted, the factor having the most effect on range error would be AOI for Methods 1A and 1B, and the factor having the most effect on range noise would be range.

4.2.2   Effect of AOI on Range Error

The dashed lines in Figure 19 to Figure 31 indicate the upper (7 mm) and lower (-7 mm) bounds of the manufacturer's specified range uncertainty. As seen in Figure 19, Methods 1A and 1B have similar trends and values. In both methods, AOI does not appear to have much of an effect on range error for AOIs ≤ 40°. However, there is a significant increase in the range error for an AOI of 60°. The trend in Figure 19 shows that the underestimation of the range increases for increasing AOI and the standard deviation of the range error increases for increasing AOI.

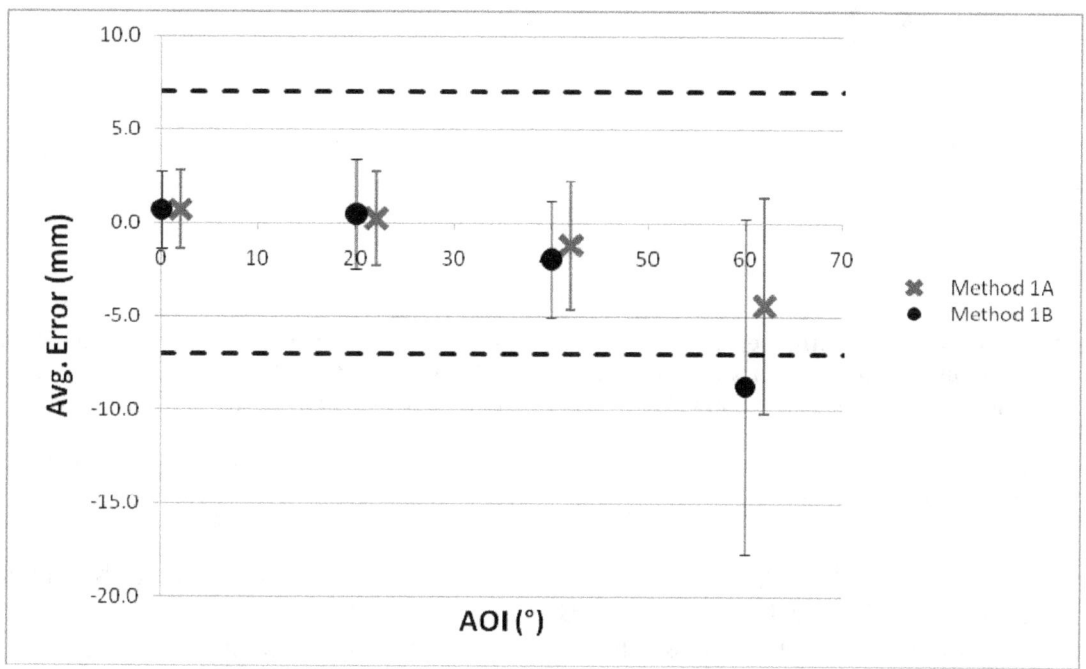

Figure 19.   Average error vs. AOI. Points represent the averages for a given AOI over all ranges, reflectivities, and azimuths. The error bars are the standard deviations of the data. Note: AOI values for the two data sets have equal values but are plotted with a slight offset for clarity.

The RMS vs. AOI, Figure 20, shows that the RMS is stable for AOI ≤ 20°, starts to increase at about an AOI of 40°, and shows a large increase at an AOI of 60°. A comparison of the two plots in this figure show that the 2 % reflective target increases the range noise and the standard deviation of the range noise.

31

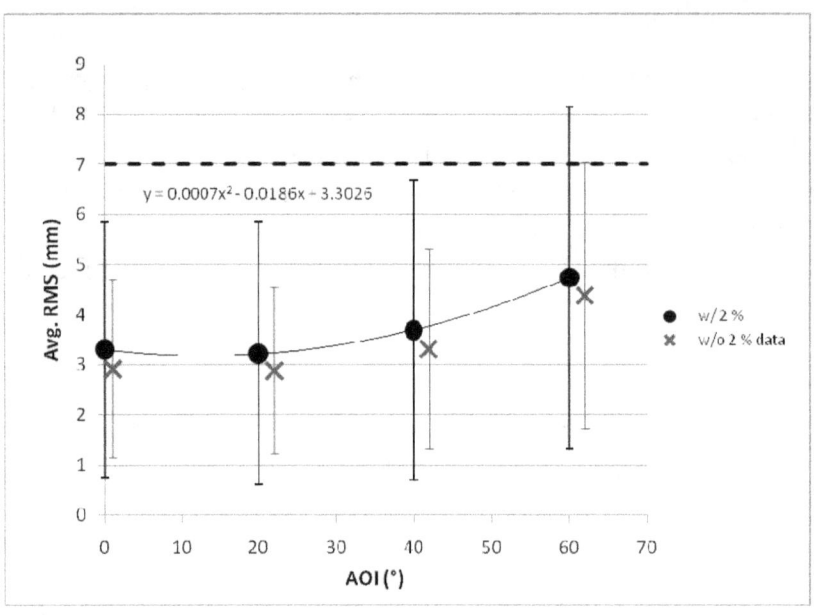

The equation shown on the graph:
$$y = 0.0007x^2 - 0.0186x - 3.3025$$

Figure 20. RMS vs. AOI. Points represent the averages for a given AOI over all ranges, reflectivities, and azimuths. The error bars represent the standard deviation of the data. Note: AOI values for the two data sets have equal values but are plotted with a slight offset for clarity.

4.2.3 Effect of Range on Range Error

Figure 21 shows the relationship between the range error and range. The curves for Methods 1A and 1B are similar in shape with the Method 1B curve lower than the curve for Method 1A. Because the curve for Method 1A crosses the x-axis, there are two groupings of the data in Figure 21: for ranges ≤ 60 m, the instrument is overestimating the range and for ranges > 60 m, the instrument is underestimating the ranges. The curve for Method 1B is below the x-axis indicating that the instrument is underestimating the range for all ranges. As mentioned in Section 2.2.6, the instrument was focused at 50 m, i.e., the measurements were more accurate at 50 m. This being the case, the data for Method 1B makes more sense as the range error is smallest at 60 m; whereas the range error is smallest at 15 m for Method 1A.

The standard deviations of the range error increase as the range increases for Method 1A. In contrast, the standard deviations of the range error decrease as the range increases for Method 1B. The behavior observed for Method 1A is more commonly expected.

32

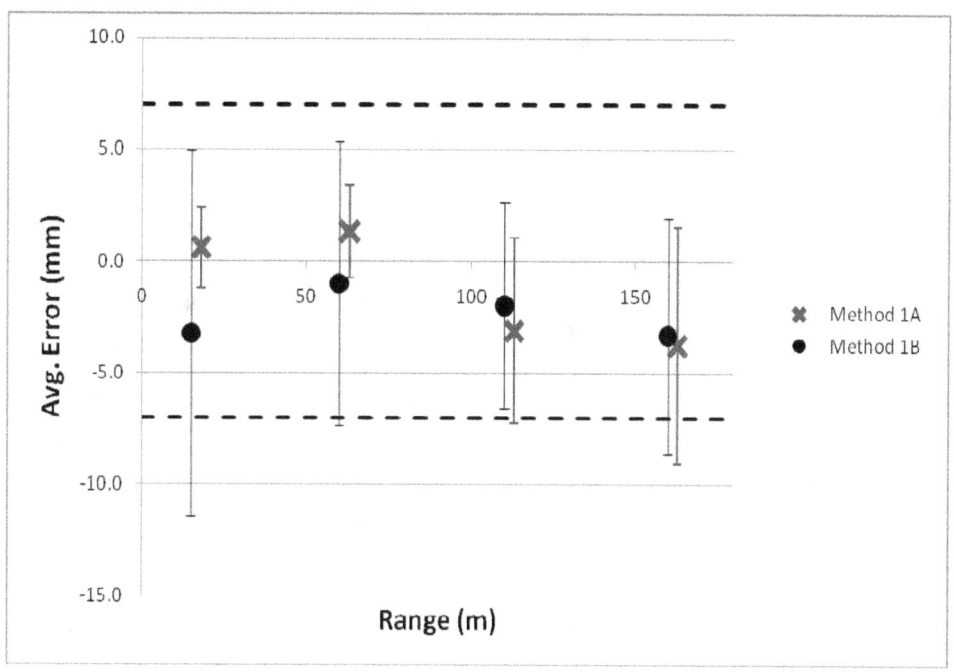

Figure 21. Error vs. range. Points represent the averages for a given range over all reflectivities, AOIs, and azimuths. There were no data for the 2 % reflective target for ranges > 60 m. The error bars are the standard deviations of the data. Note: Range values for the two data sets have equal values but are plotted with a slight offset for clarity.

The relationship between instrument noise and range is shown in Figure 22. In contrast to the range error, the 2 % reflective target significantly increases the range noise as indicated by the larger average value and much larger standard deviation. There is also a clear relationship between range noise and range: range noise increases linearly with range. The regression line shown in Figure 22 is fitted to the "w/o 2 %" data ("x" markers).

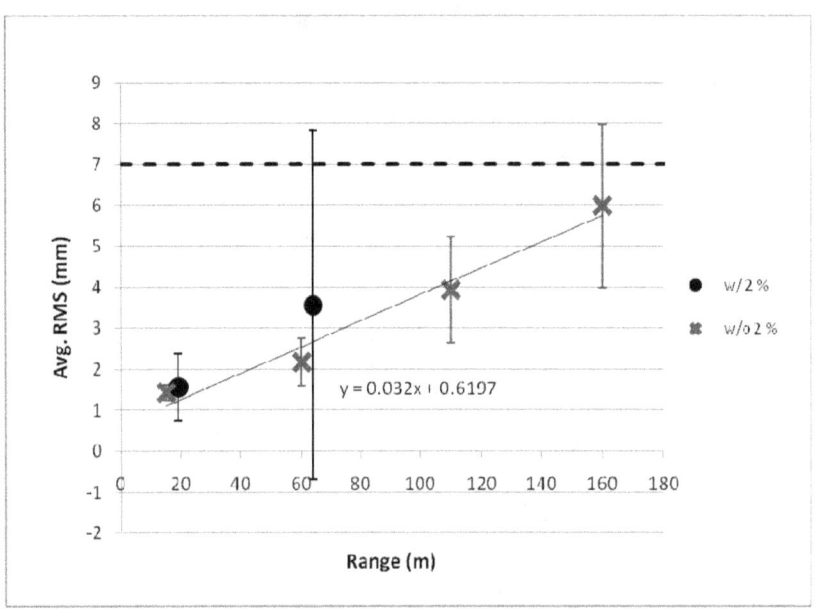

Figure 22. **RMS vs. range. Points represent the averages for a given range over all reflectivities, AOIs and azimuths. There were no data for the 2 % reflective target for ranges > 60 m. The error bars are the standard deviations of the data. Note: Range values for the two data sets have equal values but are plotted with a slight offset for clarity.**

4.2.4 Effect of Reflectivity on Range Error

Except for the 2 % reflective target, Methods 1A and 1B show similar trends - the magnitude of the range error increasing as the reflectivity decreases (Figure 23). The difference between the two methods occurs for the 2 % reflective target. The reason for the change in signs for the range error for Method 1A is unclear. The trend observed for Method 1B is what would be expected.

It is also expected that the standard deviation of the range error would be largest for the 2 % reflective target. This is the case for Method 1B where the standard deviation of the range error increases for decreasing reflectivity. However, for Method 1A, the standard deviation of the range error is largest for a 99 % reflective target.

Figure 24 shows the reduction in the number of points (i.e., return signals) for a 2 % reflective target at 60 m. The lack of points on the 2 % reflective target at 110 m is shown in Figure 25 where most of the points are from the target frame.

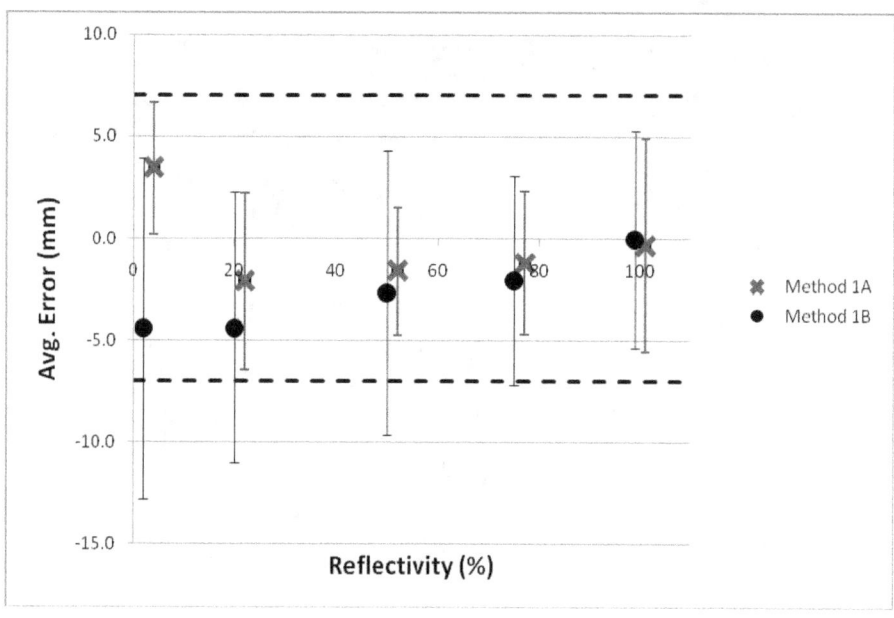

Figure 23. Average Error vs. Reflectivity. Points represent the averages at a given reflectivity over all ranges, AOIs, and azimuths. The error bars are the standard deviations of the data. Note: Reflectivity values for the two data sets have equal values but are plotted with a slight offset for clarity.

a. 75 % Reflective target

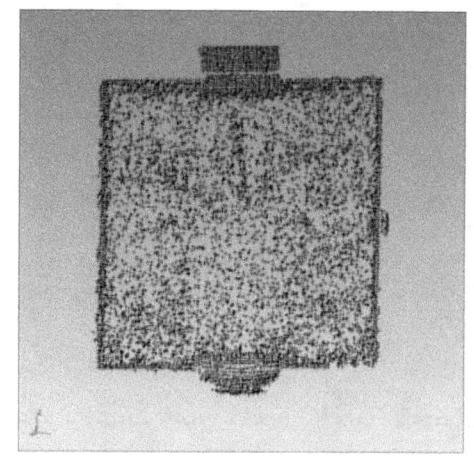

b. 2 % Reflective target

Figure 24. Difference in the number of points in the point clouds for two targets at 60 m and AOI = 0° caused by the target reflectivity.

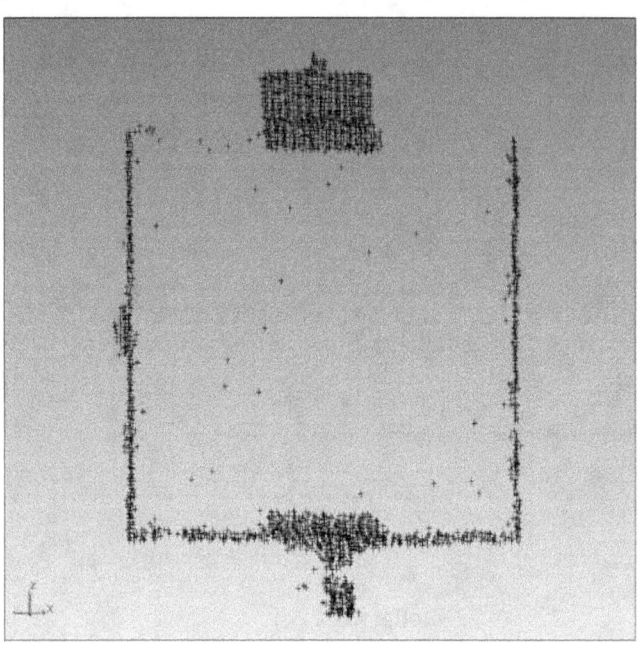

Figure 25. Point cloud for 2 % reflective target at 110 m, AOI = 0°. A minimal number of points are from the target surface with the majority of the points from the target frame and holder.

Figure 26 shows the relationship between the average RMS values and reflectivity. For reflectivities less than 50 %, the range noise increases and shows a significant increase for the 2 % reflective target. The standard deviation of the range noise also increases as the reflectivity decreases.

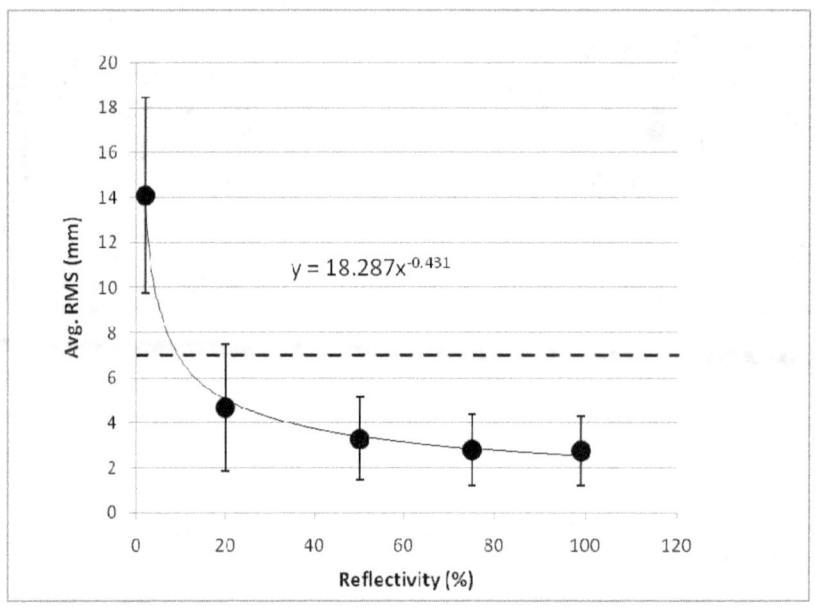

$$y = 18.287x^{-0.431}$$

Figure 26. Average RMS vs. reflectivity. Points represent the average RMS value for a given reflectivity over all AOIs, ranges and azimuths. Error bars are the standard deviations of the data.

36

4.2.5   Effect of Azimuth on Range Error

The data in Figure 19 to Figure 26 were separated by azimuth and re-plotted as Figure 27 and Figure 28. Since the data for the 2 % reflective target were only obtained for Azimuth 2, the data from the 2 % tests were omitted from Figure 27 and Figure 28. The trends for the average range errors and RMS values show that the azimuth has minimal effect on the range error and no effect on range noise. No differences between Method 1A and 1B were noted.

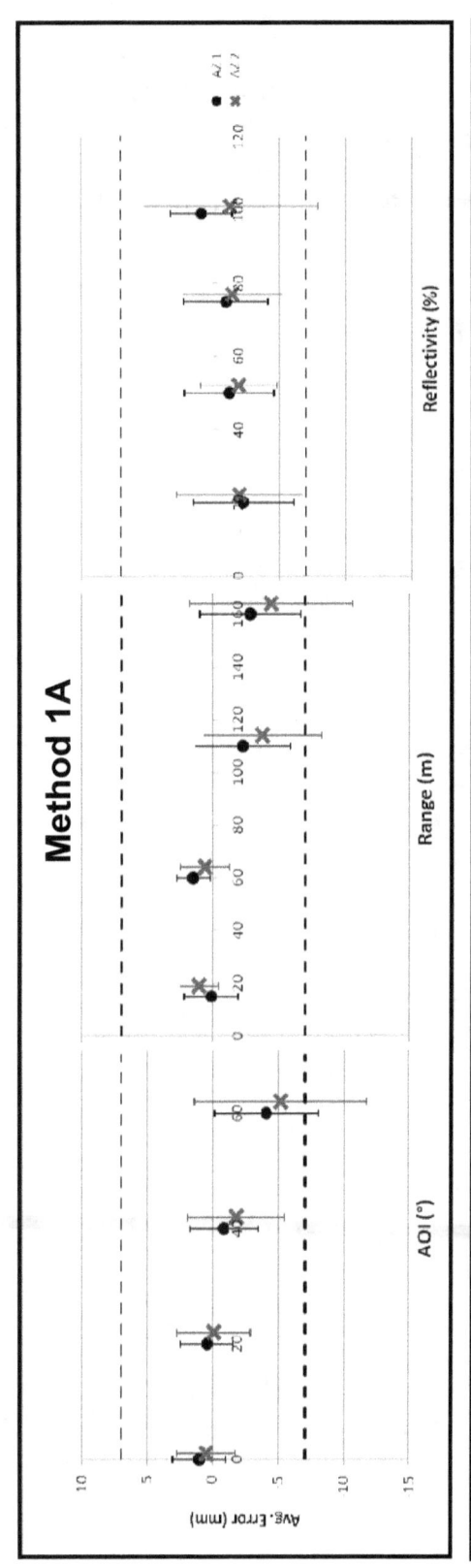

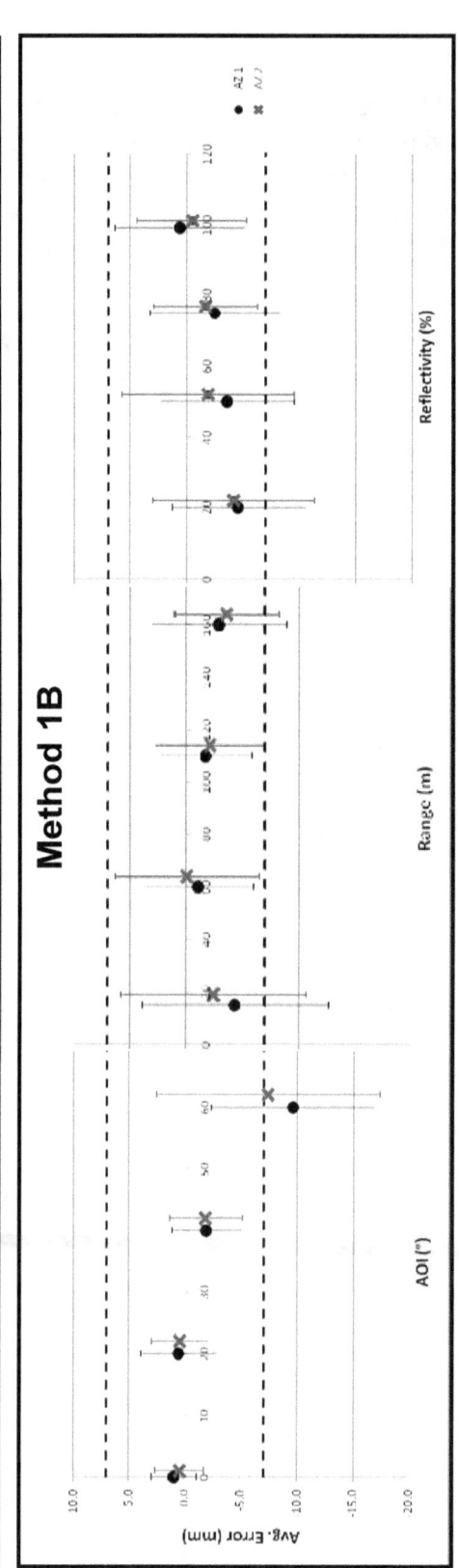

a. AOI: Points represent the average errors at a given AOI and azimuth over all ranges and reflectivity.

b. Range: Points represent the average errors for a given range and azimuth over all reflectivities and AOIs.

c. Reflectivity: Points represent the average errors for a given reflectivity and azimuth over all ranges and AOIs.

Figure 27. Error Plots - Effect of Azimuth. Error bars are the standard deviations of the data. Data for 2 % reflective target omitted. Note: AOI values for the two data sets have equal values but are plotted with a slight offset for clarity.

38

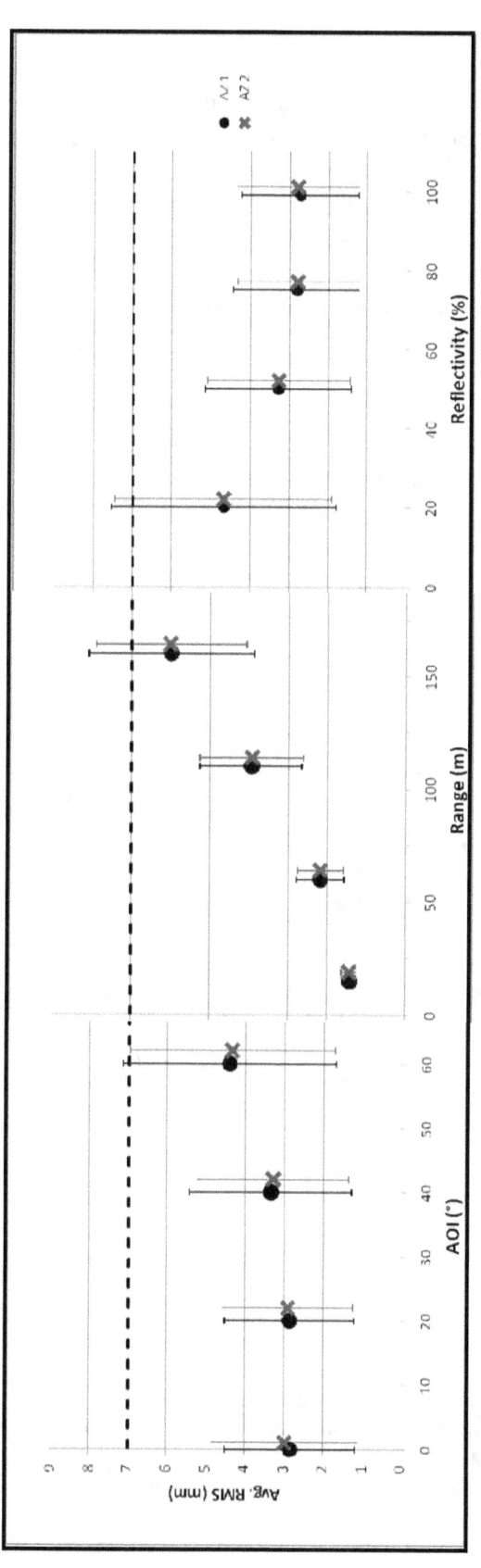

a. AOI: Points represent the average RMS values for a given AOI and azimuth over all reflectivities and ranges.

b. Range: Points represent the average RMS values for a given range and azimuth over all reflectivities and AOIs.

c. Reflectivity: Points represent the average RMS values for a given reflectivity and azimuth over all AOIs and ranges.

Figure 28. RMS Plots – Effect of Azimuth. The error bars are the standard deviations of the data. Data for 2 % reflective target omitted. Note: Range values for the two data sets have equal values but are plotted with a slight offset for clarity.

4.2.6 Reproducibility

As mentioned earlier, of the 128 tests in Test Set A, 16 of these tests were repeated on a different day. In this instance, the test procedure, operators, and environment were the same for the tests. The only difference was the time between the tests. The 16 repeat measurements were performed for Azimuth level = 2. Two criteria were used for determination of reproducibility – one based on statistics (binomial and t tests) and the second based on the manufacturer specification. For the second criterion, a pair of tests is reproducible if the difference in the range error between a pair is within ± 14 mm which simulates the situation when the uncertainties for the two ranges are at the extreme ends of the manufacturer-specified range uncertainty.

A comparison of the errors for the 16 pairs of tests is shown in Figure 29 for Method 1A and in Figure 30 for Method 1B. The differences between the average errors for the 16 pairs are shown in Figure 31 for Method 1A and in Figure 32 for Method 1B. In Figure 29 to Figure 32, within each grouping of AOI, the ranges increase from left to right. For example, for AOI = 20°, the corresponding ranges for Pairs 5, 6, 7, and 8 are (15, 60, 110, and 160) m, respectively.

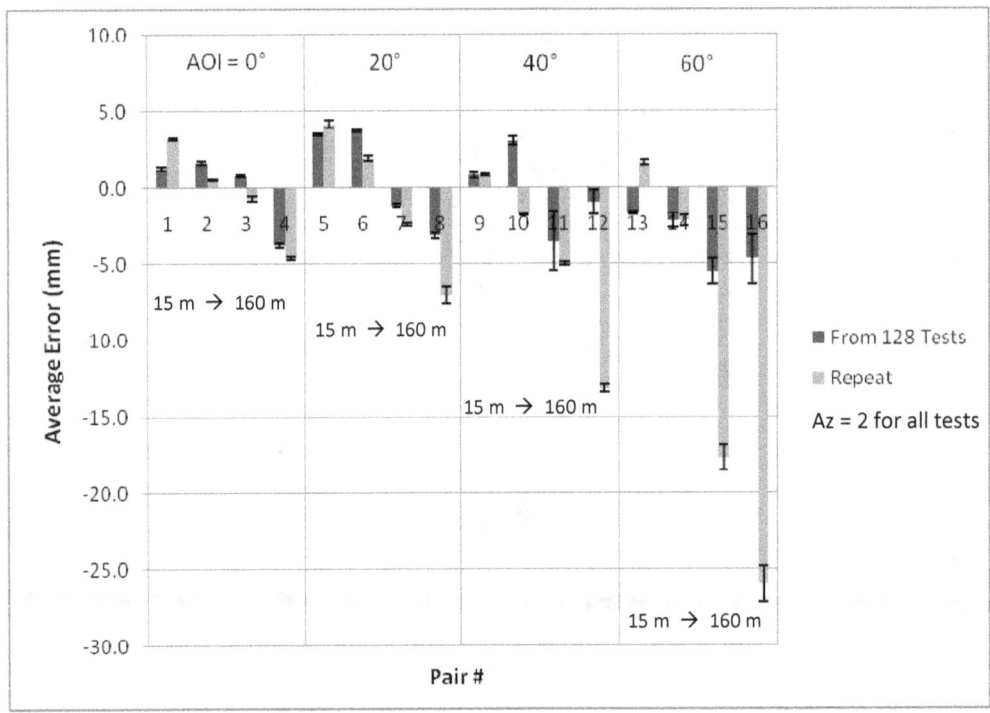

Figure 29. Method 1A: Comparison of 16 pairs of tests from Test Set A. The error bars represent the standard deviation (3 repeats) of the data. Note the degradation of reproducibility for increasing AOI.

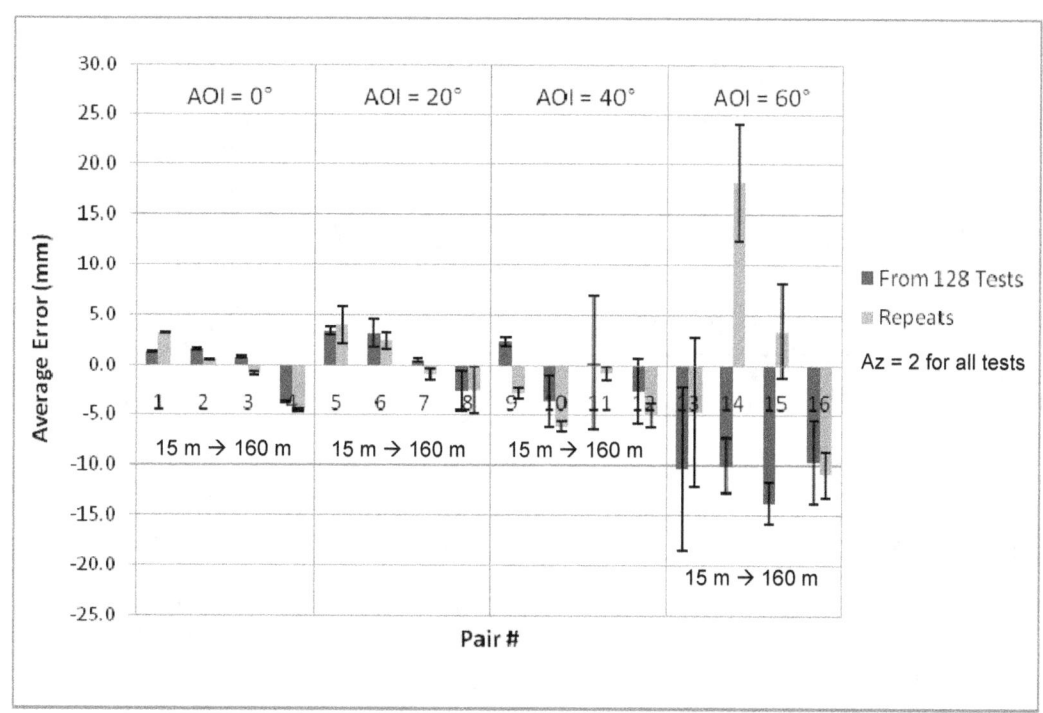

Figure 30. Method 1B: Comparison of 16 pairs of tests from Test Set A. The error bars represent the standard deviation (3 repeats) of the data. Note the degradation of reproducibility for AOI = 60°.

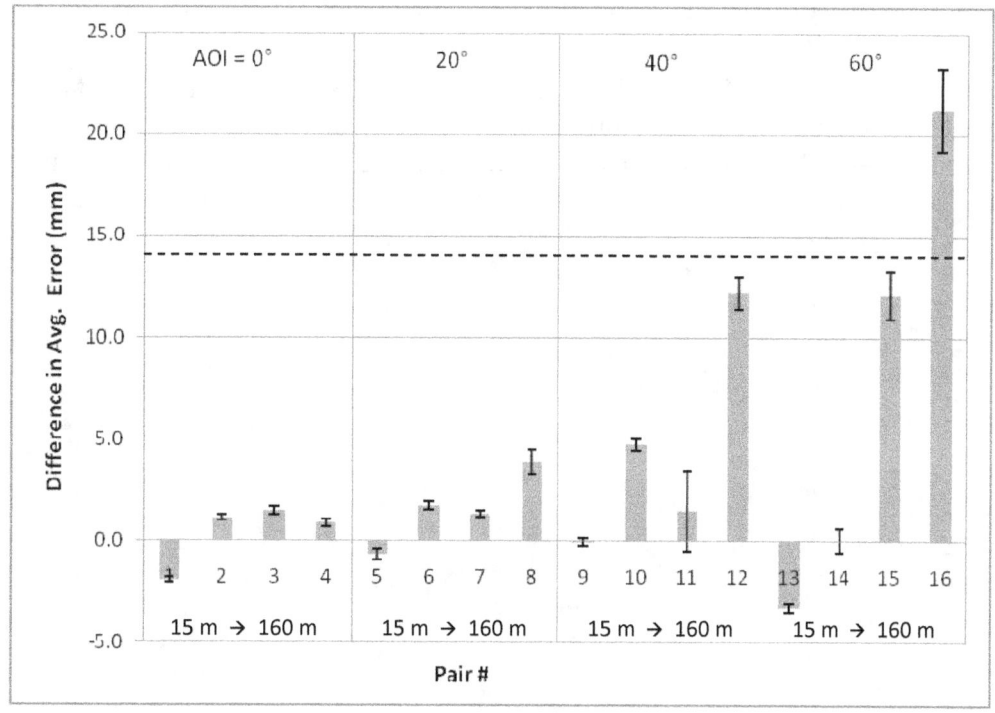

Figure 31. Method 1A: Comparison of the differences in the average error for the 16 pairs shown in Figure 29. The error bars represent the standard uncertainty. The dashed line represents twice the upper limit of the manufacturer specified range uncertainty.

41

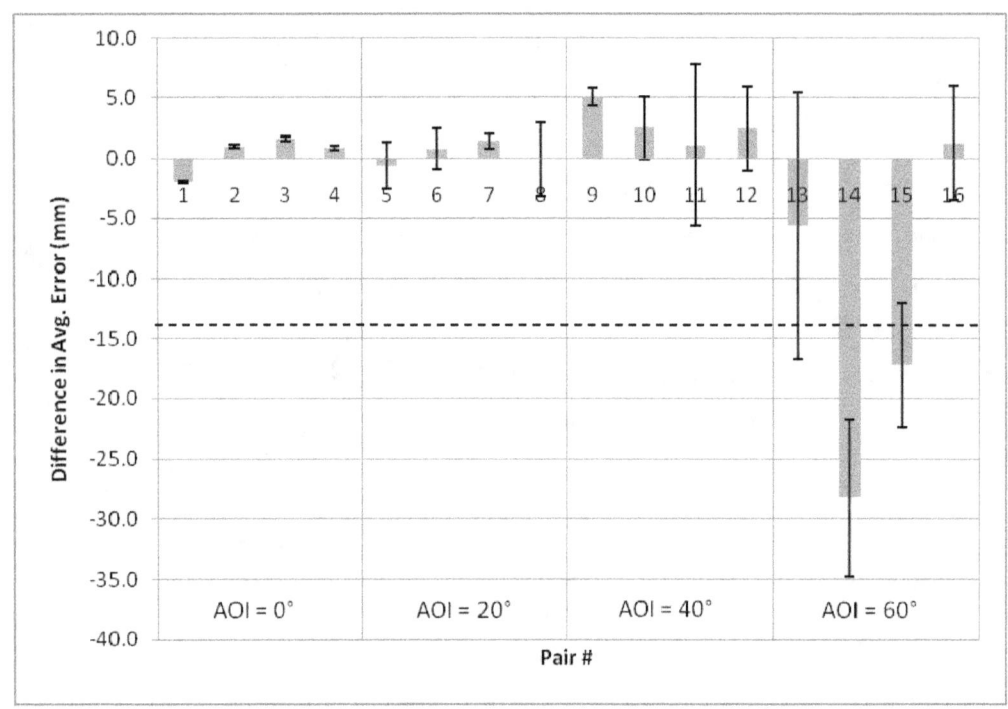

Figure 32. Method 1B: Comparison of the differences in the average error for the 16 pairs shown in Figure 30. The error bars represent the standard uncertainty. The dashed line represents twice the lower limit of the manufacturer specified range uncertainty.

The set of 16 tests from the 128 tests will be labeled as the 128-Test Set and the set of 16 repeat tests will be labeled Repeat Set. The reproducibility conclusions based on the binomial and the t tests are given in Table 3. As shown in Table 3, the tests are not reproducible if Method 1A was used to process the data; however, the tests are reproducible if Method 1B was used.

Table 3. Reproducibility Based on Binomial and t tests.

	Binomial Sign Test	t test
Method 1A, based on all 16 pairs	The range error from the 128-Test Set is greater than the error from the Repeat Set 13 out of 16 times (which falls at the 98.9 % point, which means that it has only a 2.2 % probability of occurring if in truth there was no difference between the 2 sets) → STATISTICALLY DIFFERENT	t statistic = 2.21 which falls at the 97.9 % point, which means that it (or worse) has a 4.2 % (2.1 % in each tail) probability of occurring "by chance". Since 4.2 % is less than the usual 5 % level for hypothesis testing, → STATISTICALLY DIFFERENT
Method 1B, based on all 16 pairs	The range error from the 128-Test Set is greater than the error from the Repeat Set 10 out of 16 times (which falls at the 77.3 % point, which means that it has a 45.4 % probability of occurring if in truth there was no difference between the 2 sets) → NOT STATISTICALLY DIFFERENT	t statistic = 1.05 which falls at the 84.5 % point, which means that it (or worse) has a 31 % (15.5 % in each tail) probability of occurring "by chance". Since 31 % is greater than the usual 5 % level for hypothesis testing, → NOT STATISTICALLY DIFFERENT
Method 1A, excluding pairs 12, 15, and 16	The range error from the 128-Test Set is greater than the error from the Repeat Set 10 out of 13 times (which falls at the 95.4 % point, which means that it has a 9.4 % probability of occurring if in truth there was no difference between the 2 sets) → NOT STATISTICALLY DIFFERENT	t statistic = 1.42 which falls at the 91 % point, which means that it (or worse) has a 18 % (9.0 % in each tail) probability of occurring "by chance". Since 18 % is greater than the usual 5 % level for hypothesis testing, → NOT STATISTICALLY DIFFERENT

Based on the data in Figure 31, for Method 1A the large discrepancies between the two sets occur for combinations of long range (110 m and 160 m) and high AOI (40° and 60°) – pairs 12 (160 m and 40°), 15 (110 m and 60°), and 16 (160 m and 60°). For Method 1A, if pairs 12, 15, and 16 were excluded from the analysis, the tests are reproducible (see 3rd row in Table 3).

Based on the second criteria (uncertainty of the instrument), the tests are reproducible except for AOI = 60° for both Methods 1A and 1B.

The conclusions on reproducibility are:

1. Based on statistics:
   a. Method 1B: tests are reproducible over all AOIs, ranges, and reflectivities.
   b. Method 1A: tests are reproducible for AOIs of 0° and 20° over all ranges and reflectivities.
2. Based on instrument specifications:
   a. Methods 1A and 1B: tests are reproducible except for AOI = 60°
3. The combination of longer range and high AOI (60°) reduces the chances for reproducibility. This lack of reproducibility for this combination is likely caused by the target alignment problems as any misalignments are amplified by this combination of factors.

4.3 Method of Obtaining Range to Target: Measuring Single Point on Target vs. Scanning Target

The simplest way to evaluate the range performance of a 3D imaging system is to measure the range to a target using a single measurement and to compare that value to a reference measurement. However, as discussed in Section 2.1.2, the development of a protocol to evaluate the range performance of a 3D imaging system may have to involve scanning a target. The issue with scanning a target is that the process of scanning introduces errors from the angular movement of the instrument; thus, the resulting error would not be a pure range error. However, since the errors from the angular encoders are generally much smaller than the error from the range measurement mechanism and since the angular movement involved in scanning a target is relatively small, the range error as obtained by scanning a target can be taken to be equivalent to the pure range error.

The comparison of results obtained by scanning the target, Methods 1A and 1B, with those from Method 2 will highlight any differences in the results due to the method used to obtain the range measurement. These differences, if any, will help in the development of test protocols to evaluate the ranging performance of 3D imaging systems. Figure 33 offers some clarification between Methods 1A and 1B and Method 2.

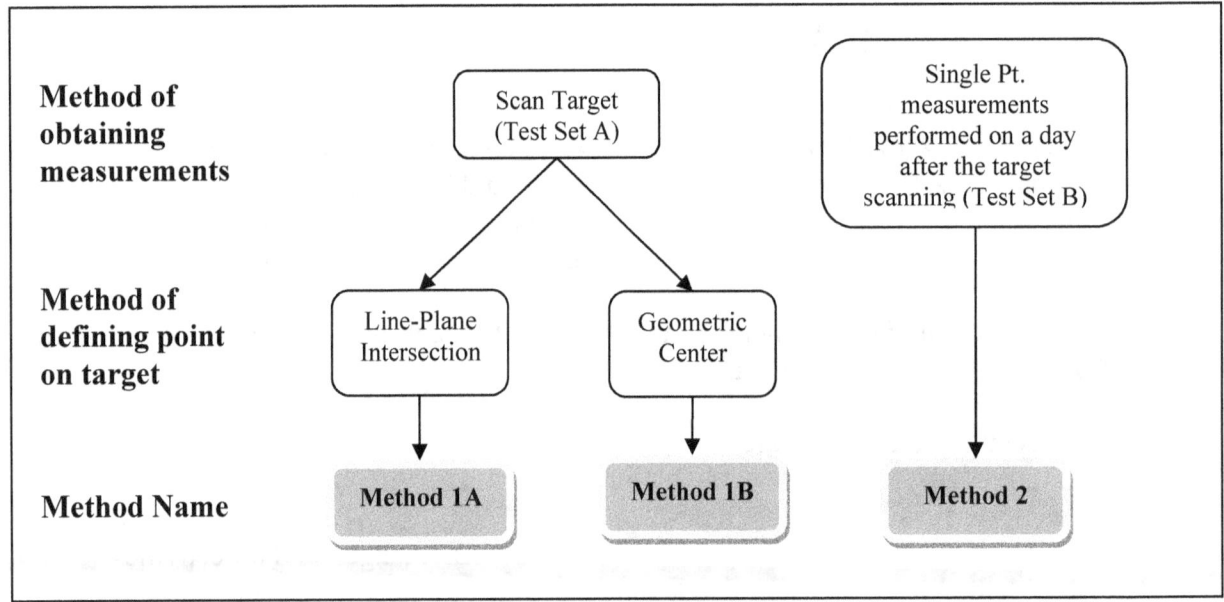

Figure 33.   Schematic Describing the Different Methods.

4.3.1   Analysis Based on Range Errors

It is re-iterated that the ranges for Methods 1A and 1B were from Test Set A and those for Method 2 were from Test Set B.

Figure 34[g] shows an overview of the data for the two methods of obtaining measurements: Method 1A vs. Method 2 and for Method 1B vs. Method 2. For Method 1A, Figure 34 (top plot) shows that the Method 2 error exceeds the Method 1A error in 21 (# of "heads") out of the 32 cases. If the two methods were equivalent, then one would expect a 16/16 split (i.e., that the Method 2 error would exceed the Method 1A error half of the time) on average. Based on the binomial sign test (referred to as the binomial test henceforth), the observed 21/11 (or 11/21) split has a 11.0 % (5.5 % in each tail) probability of occurring, which is not significant at the 5 % level. The advantage of the binomial test is that it is a conservative test which does not make assumptions about the underlying distribution; the disadvantage is that it does not incorporate information about the magnitude of the observed differences and hence may at times be less sensitive to small method differences when they do in fact exist.

---

[g] It is NIST's policy to provide uncertainties associated with reported measurements. However, no uncertainties are given for the values in Figure 34 because the purpose of this figure is to show obvious trends (e.g., Method 2 error is greater than Method 1A error for the majority of cases). Additionally, these values with their associated uncertainties are given in Figure 35 to Figure 37.

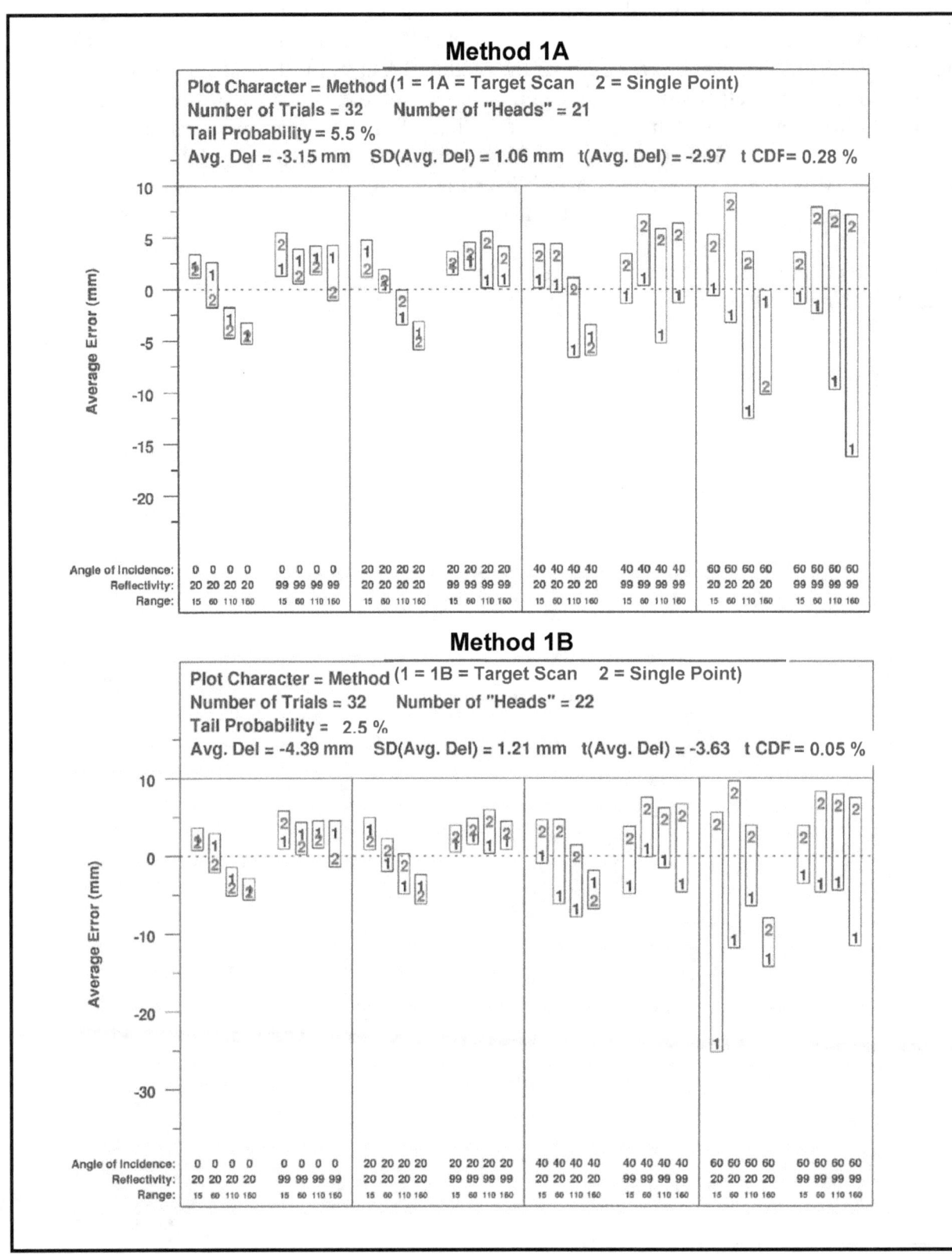

Figure 34.   Average error plots to compare Method 1A (target scan) and Method 2 (single point measurements) – top plot and Method 1B (target scan) and Method 2 (single point measurements) – bottom plot.

The alternative t test does make use of such magnitude information; the t statistic (= average del / SD{average del})[h] has the value of (-3.15/1.06) = -2.97 has a 0.56 % (2 x 0.28 %) chance of occurring (by chance) and hence is significant at both the usual 5 % level and even the 1 % level. Hence, the two methods, 1A and 2, are statistically different, and on average, the Method 2 error exceeds the Method 1A error by 3.2 mm.

The differences between the Methods 1A and 2 and Methods 1B and 2 are more pronounced as the AOI increases. As AOI increases, the average difference between Method 2 and Method 1A increases, with differences of -0.9 mm[i], +0.8 mm, +4.1 mm, and +8.7 mm for AOIs of 0°, 20°, 40°, and 60°, respectively. The large difference in error between the two methods at higher AOIs, especially at AOI = 60°, may be caused by misalignment of the target (e.g. off center, tilted target). This points again to the importance of target set up.

Since the differences between Methods 1A and 2 are dependent on AOI, the results from the binomial and t tests for the four levels of AOI and employing the binomial and t tests at the 5 % level for Method 1A are given in Table 4. As seen in Figure 34, the results for Method 1A and 1B are similar and the results for Method 1B (Table 5) reflect this. For Method 1B, the conclusions for significance between target scanning and single point measurement are the same as those for Method 1A.

---

[h] average del = average difference between the two methods
SD(average del) = standard deviation of the average del
[i] A negative value indicates that the Method 1A or 1B error was higher than the Method 2 error. A positive value indicates that the Method 2 error was higher than the Method 1A or 1B error.

Table 4. Comparison of Method 1A (Line-Plane Intersection, Target Scan) and Method 2 (Single Point Measurement).

AOI (°)	Binomial Sign Test	t test
0	Method 2 exceeds Method 1A 1 out of 8 times (which falls at the 3.5 % point, which means that it has only a 7.0 % probability of occurring if in truth there was no difference in the 2 methods) → **NOT STATISTICALLY DIFFERENT**	t statistic = -1.55 which falls at the 7.9 % point, which means that it (or worse) has a 15.8 % (7.9 % in each tail) probability of occurring "by chance". Since 15.8 % exceeds the usual 5 % level for hypothesis testing, → **NOT STATISTICALLY DIFFERENT**
20	Method 2 exceeds Method 1A in 6 out of the 8 cases (which falls at the 85.5 % point, which means that it has only a 29.0 % probability of occurring if in truth there was no difference in the 2 methods) → **NOT STATISTICALLY DIFFERENT**	t statistic = 1.34 which falls at the 89.1 % point, which means that it (or worse) has a 21.8 % (10.9 % in each tail) probability of occurring "by chance". Since 21.8 % exceeds the usual 5 % level for hypothesis testing, → **NOT STATISTICALLY DIFFERENT**
40	Method 2 exceeds Method 1A in 7 out of the 8 cases (which falls at the 96.5 % point, which means that it has only a 7.0 % probability of occurring if in truth there was no difference in the 2 methods) (statistically different) → **NOT STATISTICALLY DIFFERENT**	t statistic = 3.78 which falls a the 99.7 % point, which means that it (or worse) has 0.6 % (0.3 % in each tail) probability of occurring "by chance". Since 0.6 % is less than the usual 5 % level and even the 1 % for hypothesis testing, → **STATISTICALLY DIFFERENT**
60	Method 2 exceeds Method 1A in 7 out of 8 cases (which falls at the 96.5 % point, which means that it has only a 7.0 % probability of occurring if in truth there was no difference in the 2 methods) → **NOT STATISTICALLY DIFFERENT**	t statistic = 2.68 which falls at the 98.6 % point, which means that it (or worse) has 2.8 % (1.4 % in each tail) probability of occurring "by chance". Since 2.8 % is less than the usual 5 % level for hypothesis testing, → **STATISTICALLY DIFFERENT**

AOI (°)	Binomial Sign Test	t test
0	Method 2 exceeds Method 1B 1 out of 8 times (which falls at the 3.5 % point, which means that it has only a 7.0 % probability of occurring if in truth there was no difference in the 2 methods) → NOT STATISTICALLY DIFFERENT	t statistic = -1.59 which falls at the 7.5 % point, which means that it (or worse) has a 15.0 % (7.5 % in each tail) probability of occurring "by chance". Since 15.0 % exceeds the usual 5 % level for hypothesis testing, → NOT STATISTICALLY DIFFERENT
20	Method 2 exceeds Method 1B in 6 out of the 8 cases (which falls at the 85.5 % point, which means that it has only a 29.0 % probability of occurring if in truth there was no difference in the 2 methods) → NOT STATISTICALLY DIFFERENT	t statistic = 1.56 which falls at the 92.2 % point, which means that it (or worse) has a 15.6 % (7.8 % in each tail) probability of occurring "by chance". Since 15.6 % exceeds the usual 5 % level for hypothesis testing, → NOT STATISTICALLY DIFFERENT
40	Method 2 exceeds Method 1B in 7 out of the 8 cases (which falls at the 96.5 % point, which means that it has only a 7.0 % probability of occurring if in truth there was no difference in the 2 methods) (statistically different) → NOT STATISTICALLY DIFFERENT	t statistic = 4.10 which falls a the 99.8 % point, which means that it (or worse) has 0.4 % (0.2 % in each tail) probability of occurring "by chance". Since 0.4 % is less than the usual 5 % level and even the 1 % for hypothesis testing, → STATISTICALLY DIFFERENT
60	Method 2 exceeds Method 1B in 8 out of 8 cases (which falls at the 99.6 % point, which means that it has only a 0.8 % probability of occurring if in truth there was no difference in the 2 methods) → STATISTICALLY DIFFERENT, METHOD 2 EXCEEDING METHOD 1B	t statistic = 4.30 which falls at the 99.9 % point, which means that it (or worse) has 0.2 % (0.1 % in each tail) probability of occurring "by chance". Since 0.2 % is less than the usual 5 % level for hypothesis testing and even the 1 % for hypothesis testing, → STATISTICALLY DIFFERENT

In summary, it is seen that the significance of Methods 1A and 1B (target scan) and Method 2 (single point measurement) is strongly dependent on AOI,

- For AOI = 0° and 20°, the binomial test and t test indicate that both that target scanning and single point measurement are not statistically different.
- For AOI = 40°, for Methods 1A and 1B, the binomial indicates no significance while the t test indicates statistical difference between the target scanning and single point measurement.
- For AOI = 60°,
  - Method 1A – binomial test indicates no significance while the t test indicates statistical difference between the target scanning and single point measurement.
  - Method 1B - both the binomial and t test indicate statistical difference between the target scanning and single point measurement.

Figure 35, Figure 36, and Figure 37 show plots of the data in Figure 34 as a function of AOI, range, and reflectivity, respectively. The data in Figure 34 to Figure 37 were averaged for reflectivities of 20 % and 99 % and only Azimuth 2 data was used since the single point measurements (Method 2) were only obtained for this azimuth and these two reflectivities. In Figure 35 to Figure 37, "Single Pts" (Method 2) data are from Test Set B and "Target Scan" (Method 1A and 1B) data are from Test Set A.

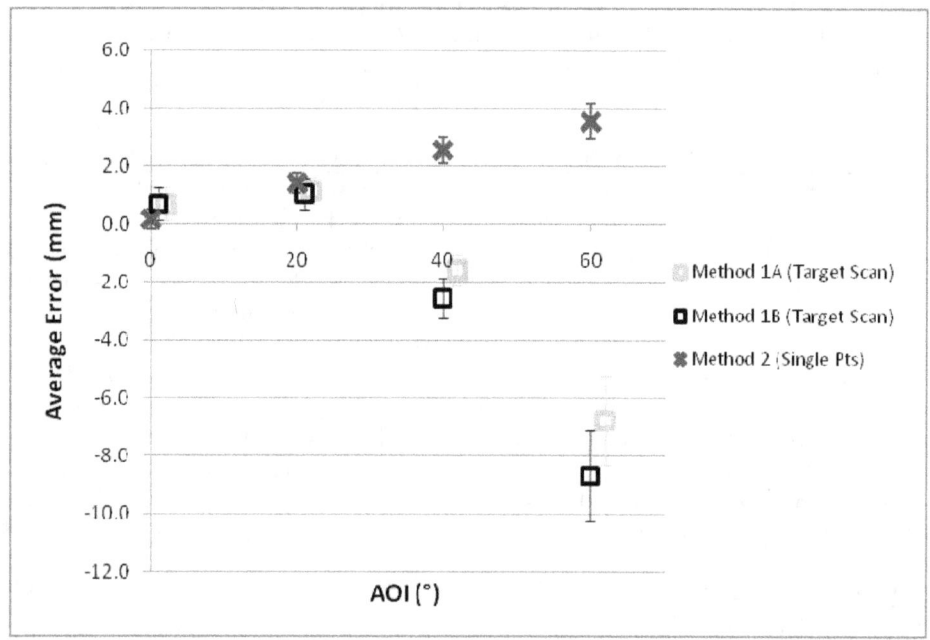

Figure 35. **Comparison of Methods 1A, 1B, and 2. Points represent the averages for a given AOI over all ranges and reflectivities of 20 % and 99 %. For Methods 1A and 1B, only the data for Az = Level 2 were used as the single point measurements were obtained for Az = Level 2 only. The error bars are the standard deviations of the mean. Note: AOI values for all data sets have equal values but are plotted with a slight offset for clarity.**

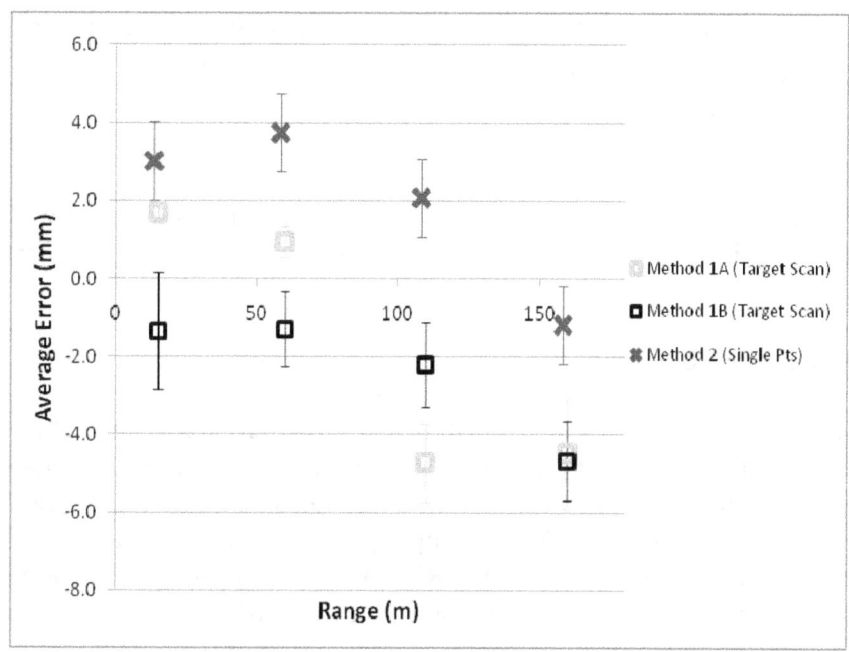

Figure 36.   Comparison of Methods 1A, 1B, and 2 based on Range. Points represent the averages for a given range over all AOIs and reflectivities of 20 % and 99 %. For Methods 1A and 1B, only the data for AZ 2 were used as the single point measurements were obtained for AZ 2 only. The error bars are the standard deviations of the mean.

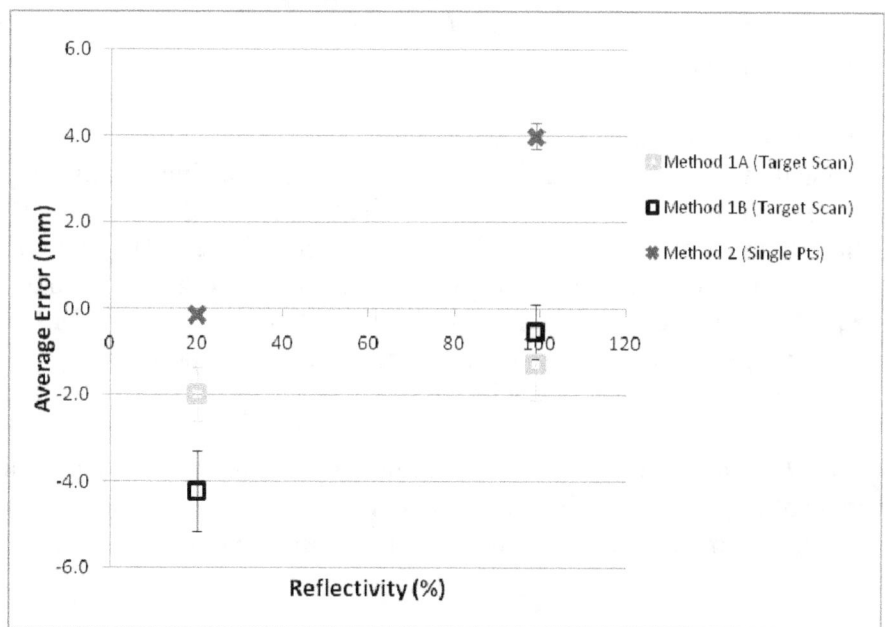

Figure 37.   Comparison of Methods 1A, 1B, and 2 based on Reflectivity. Points represent the averages for a given reflectivity over all ranges and AOIs. For Methods 1A and 1B, only the data for AZ 2 were used as the single point measurements were obtained for AZ 2 only. The error bars are the standard deviations of the mean.

51

The results shown in Figure 35 illustrate the conclusions drawn from the statistical tests: target scanning and single point measurement yield similar results for AOIs of 0° and 20° and different results for AOIs of 40° and 60°.

In Figure 36 and Figure 37, the data from target scanning and single point measurement yield similar trends but have an offset. Of the two methods used to define the point on the target, the geometric center method, 1B, yields results that are more similar to the single point measurement.

4.3.2   Analysis Based on Range Noise

It is common practice to use the standard deviation of individual measurements as the noise for that particular measurand. For the single point measurements, the standard deviation of single range measurements is, therefore, the noise of the range measurement or range noise. However, some 3D imaging systems are not able to take single point measurements and another value is needed to represent instrument noise. Often the RMS value from fitting a geometric shape is used. The question then is "Is the RMS (of plane fit for the experiments in this report) equivalent to the standard deviation of single point measurements?"

The data for the analysis in this section consisted of the:
1. 105 comparisons between the standard deviations of single point measurements (Test Set A) and the RMS values for the target scans (Test Set A). The single point measurements were acquired immediately after the target scans were completed.
2. 32 comparisons between the standard deviations of single point measurements (Test Set B) with the RMS values for target scans (Test Set A). The single point measurements were made on a different day.

Figure 38[j] shows that the RMS value is higher than the standard deviation (Stdev) value 59 out of 61 times -- an occurrence which has a 0.0 % probability of occurring which is significant, i.e., the two representations of range noise are different. The t statistic = 8.02 has a 0.0 % chance of occurring (by chance) and is significant at both the 5 % level and even the 1 % level. The difference between the RMS and Stdev values is clearly evident in Figure 39. In all cases, the RMS values are greater than the standard deviation values. On average, the RMS of the plane fit is 1.6 mm ($s = 0.2$ mm) greater than the standard deviation of single point measurements. The average ratio of the RMS to the Stdev value is about 2.

As noted in Section 4.2.3 and seen from Figure 38, there is a clear dependence of range noise on range. The average of the differences between the RMS and Stdev values are (0.4, 0.8, 2.3, and 3.5) mm ($s = 0.2$ mm, 0.6 mm, 1.3 mm, 2.3 mm for ranges of 15 m, 60 m, 110 m, and 160 m, respectively).

---

[j] It is NIST's policy to provide uncertainties associated with reported measurements. However, no uncertainties are given for the values in Figure 38 because the purpose of this figure is to show obvious trends (e.g., RMS values are greater than Stdev values for the majority of cases). Additionally, these values with their associated uncertainties are shown in Figure 39.

From Figure 39, the dependence of the RMS and standard deviation on each of the four factors (AOI, range, reflectivity, and azimuth) is more pronounced for the RMS value in every case. That is, for a given factor, the difference between the maximum and minimum values, Δ, is always greater for the RMS values than for the standard deviation values. Again, target alignment problems could contribute to this effect for higher AOIs and longer ranges but it would not explain this effect for reflectivity and azimuths. However, for the azimuth, it should be noted that the $\Delta_{RMS} = 0.5$ mm and $\Delta_{Stdev} = 0.3$ mm which is a very small difference.

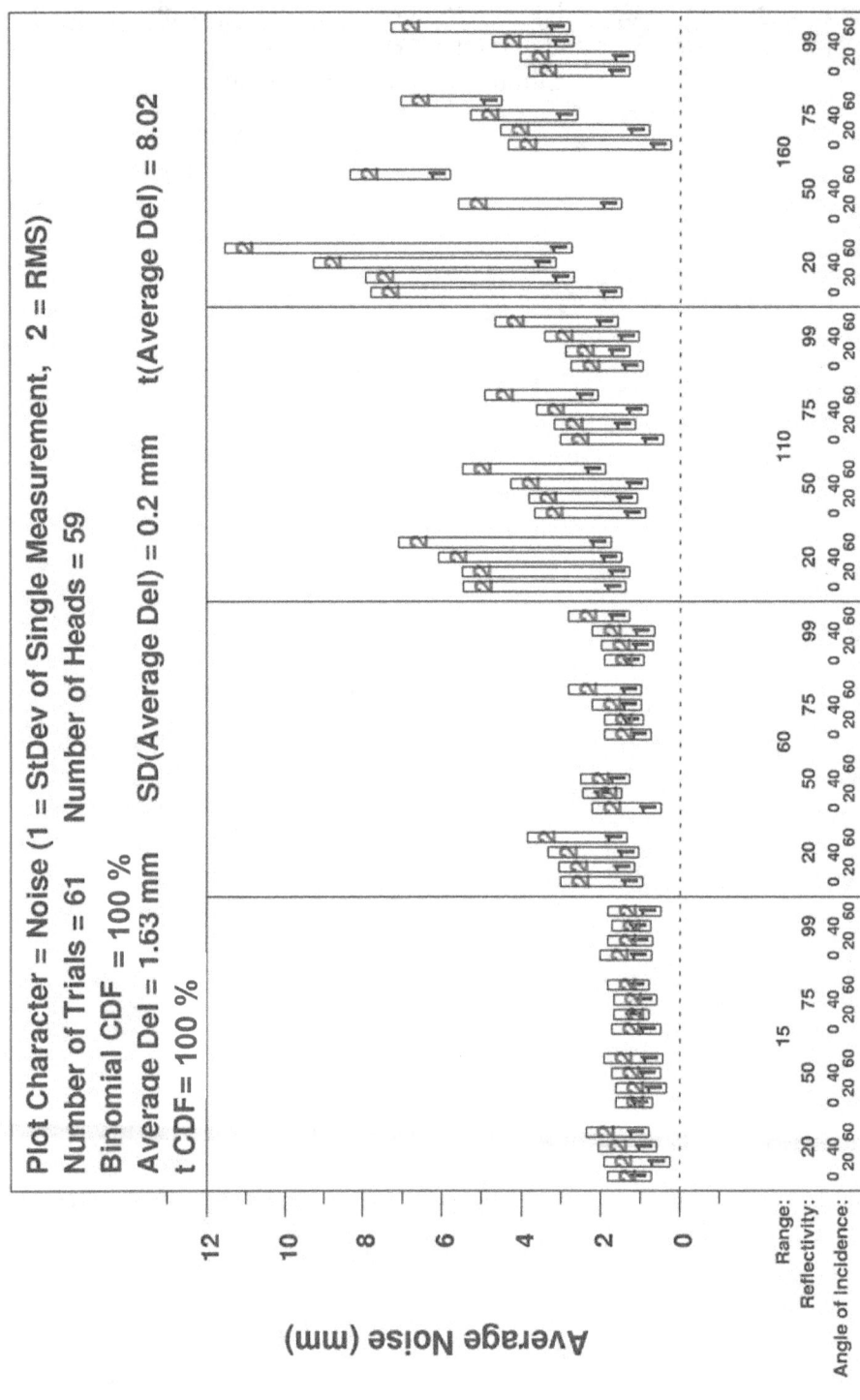

Figure 38.   Average error plots to compare the RMS of plane fit and standard deviation of single measurements.

54

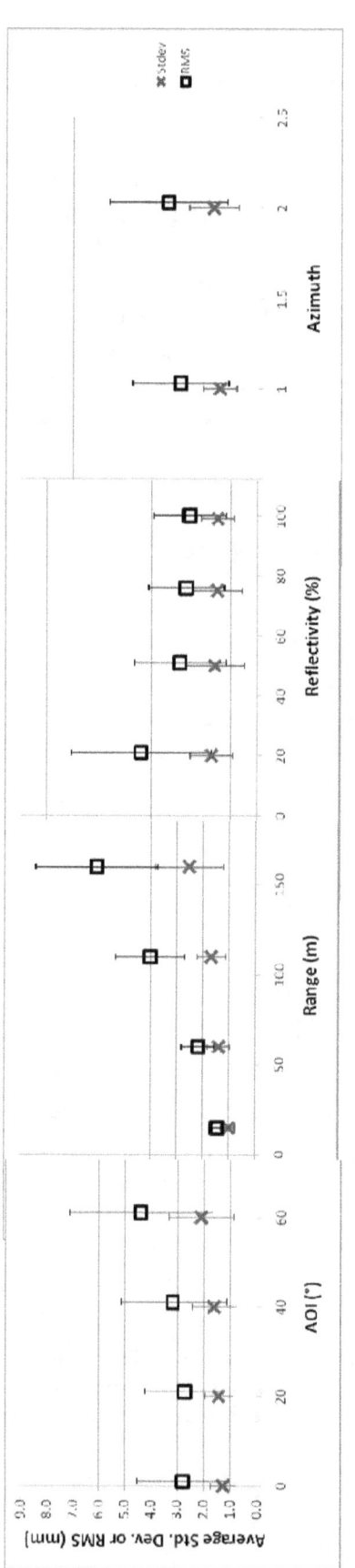

a. AOI: Points represent the averages for a given AOI over all ranges, reflectivities, and azimuths.

b. Range: Points represent the averages for a given range over all AOIs, reflectivities, and azimuths.

c. Reflectivity: Points represent the averages for a given reflectivity over all ranges, AOIs, and azimuths.

d. Azimuth: Points represent the averages for a given azimuth over all ranges, AOIs, and reflectivities.

Figure 39. Noise plots – RMS vs. Standard Deviation. The error bars are the standard deviations of the data. Note: X-axis values for the two data sets have equal values but are plotted with a slight offset for clarity.

4.4 Effect of Target Type, Planar (Test Sets A and B) vs. Spherical (Test Set C), on Range Error

The average range error is plotted for various ranges in Figure 40 for a spherical target (Target 3) and a planar target – both single point measurement (Target 2, Method 2) and target scanning (Target 1, Method 1A). The reflectivity of the spherical target was about 51 %. Therefore, for the target scanning (planar target), the data for an AOI of 0° and reflectivity of 50 % were used in Figure 40. For the single point measurement (Method 2), the data for an AOI of 0° and reflectivities of 20 % and 99 % were used as there were no data for single point measurements for 50 % reflectivity. The plot showing the comparison between Target 1 (Method 1B), Target 2 (Method 2) and Target 3 is very similar to that shown in Figure 40 and is not included in this report.

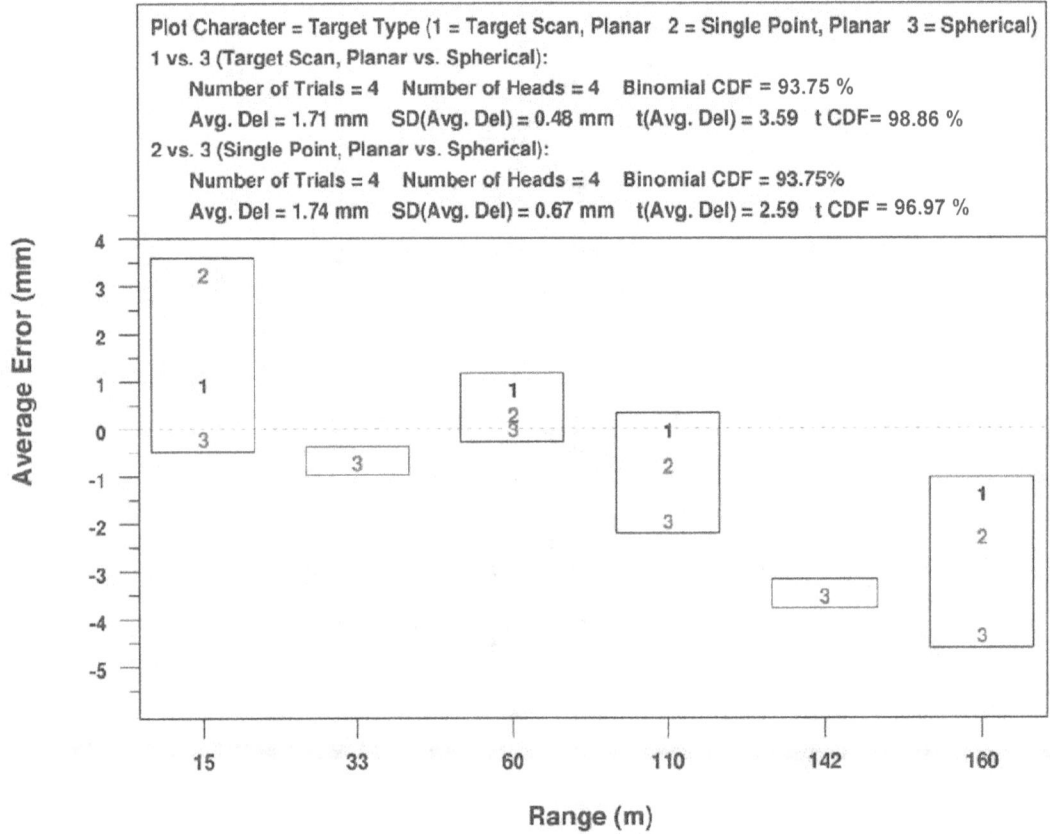

Figure 40. Error vs. Distance for different target types (planar vs. spherical). The standard deviation of the means are not shown for purposes of clarity and have the following values: Target 1 = 0.1 mm, Target 2 - varies from 0.4 mm to 0.7 mm, Target 3 - varies from 0.0 mm to 0.2 mm.

The conclusions on target equivalence based on the binomial and t tests are given in Table 6. The binomial test, though informative and suggestive with four "successes" out of four trials, is nonetheless conservative. With only four trials, it is impossible for the binomial test to yield "significant" as a conclusion at the 5 % level. Therefore, based on the t test, the Target 2 (planar target, single point measurement) is equivalent to Target 3 (spherical). However, on average, the Target 1 error is 1.7 mm greater than the Target 3 error. From an "application" perspective, when compared to the manufacturer-specified instrument uncertainty (7 mm), a 1.7 mm value could be considered not significant, i.e., the targets are equivalent.

Table 6. Binomial and t Test Results for Equivalence of a Planar and Spherical Target.

	Binomial test	t test
Target 1 (Method 1A) vs. Target 3	Target 1 exceeds Target 3 4 out of 4 times (which falls at the 93.75 % point, which means that it has only a12.5 % probability of occurring if in truth there was no difference in the 2 methods) → **NOT STATISTICALLY SIGNIFICANT**[*]	t statistic = 3.59 which falls at the 98.9 % point, which means that it (or worse) has a 2.2 % (1.1 % in each tail) probability of occurring "by chance". Since 2.2 % is less than the usual 5 % level for hypothesis testing, → **STATISTICALLY DIFFERENT**
Target 1 (Method 1B) vs. Target 3	Target 1 exceeds Target 3 4 out of 4 times (which falls at the 93.75 % point, which means that it has only a 12.5 % probability of occurring if in truth there was no difference in the 2 methods) → **NOT STATISTICALLY SIGNIFICANT**[*]	t statistic = 3.76 which falls at the 99.0 % point, which means that it (or worse) has a 2.0 % (1.0 % in each tail) probability of occurring "by chance". Since 2.0 % is less than the usual 5 % level for hypothesis testing, → **STATISTICALLY DIFFERENT**
Target 2 (Method 2) vs. Target 3	Target 2 exceeds Target 3 4 out of 4 times (which falls at the 93.75 % point, which means that it has only a 12.5 % probability of occurring if in truth there was no difference in the 2 methods) → **NOT STATISTICALLY SIGNIFICANT**[*]	t statistic = 2.59 which falls at the 97.0 % point, which means that it (or worse) has a 6.0 % (3.0 % in each tail) probability of occurring "by chance". Since 6.0 % is less than the usual 5 % level for hypothesis testing, → **NOT STATISTICALLY DIFFERENT**
[*] With only four trials, it is impossible for the binomial test to yield "significant" as a conclusion at the 5 % level.		

The use of a spherical target would make the test setup easier (e.g., alignment of target with respect to axis of rotation), the post-processing of the data much simpler (i.e., there is no need to determine which point on the plane to use), and it would reduce the sources of potential error in the measurement due to target misalignment. The disadvantages of using a spherical target are the cost to manufacture a sphere of sufficient size and sphericity required for the tests, the inability to evaluate the range performance as a function of AOI, and the practicality of manufacturing spheres with different reflectivities (since coatings can easily be damaged).

4.5   Evaluation of the Test Procedure

In terms of practicality, the test procedure used in these experiments is a viable procedure. A set of 16 tests can be easily performed in an 8 h day and a set of 32 tests can also be performed in one day (10 h to 12 h). These times are only for the actual conduct of the tests and do not include set up time – e.g., positioning and leveling of the stands, alignment of the targets.

Other observations and issues about the test procedure include:
- Planar target alignment is an important issue. This procedure is critical and is one of the more time consuming tasks. To minimize errors due to misalignment and to reduce the required time for set up, potential solutions include:
  - Re-design of the target and/or target holder.
  - Another method to post-process the data. As mentioned in Section 3, a key issue in the test procedure was defining a point on the target plane. An alternative procedure which eliminates this need involves "force fitting" the two target planes to make them parallel (i.e., when fitting the two planes, a constraint would be that the fitted planes must be parallel) which then makes the determination of the distance between the two planes trivial. The advantage of this procedure is that it does not require that the target centers be aligned perfectly which makes the experimental set up easier.
- A more direct method of obtaining the ground truth measurements. This may involve a re-design of the target and/or target holder.
- Availability of open space > 150 m in length.

4.5.1   Method of Defining Point on Target: Method 1A vs. Method 1B

In terms of range error, the two methods used to define a point on the target, Method 1A, the intersection of a line and a plane, and Method 1B, geometric center, produced similar results. The two instances where the results were dependent on method were reflectivity – see Section 4.2.4, Figure 23 and Section 4.3.1, Figure 37. In the first instance, the discrepancy only occurred for the 2 % reflective target. In the second instance, when trying to determine if target scanning would yield similar results to single point measurements, the trend for reflectivity vs. average error for single point measurements was better approximated when Method 1B was used.

The method made a difference when determining reproducibility of the tests. Based on statistical tests, Method 1A would not be reproducible while Method 1B would be considered reproducible.

Of the two methods, Method 1B is more straightforward and easier to implement than Method 1A, and the results using Method 1B seem to produce results that are slightly more rational or expected.

4.5.2 Geometric Center vs. Centroid

As mentioned in Section 3, the use of the centroid was discussed as a potential method of defining a point on the target, but the use of this method was discarded. However, since the data was available, comparisons of the locations of the geometric centers and centroids of the points in Subset 4 (Section 3.1) were made.

Out of the 446 scans, the distance between the geometric centers and centroids was less than 2 mm for 89 % of the scans. For the scans where the distance was greater than 2 mm and for reflectivities greater than 2 %, the instance where the distance was greater than 2 mm only occurred for one or two out of the three scans for a given test. For example, for Test N (where N = 1 to 149), the distance between the geometric center and the centroid might have been greater than 2 mm for scan 2 and was less than 2 mm for scans 1 and 3. In contrast, for the 2 % reflective target, the distance would have exceeded 2 mm for all three scans for a given test in the majority of cases. There were no discernible trends between the four factors (range, reflectivity, AOI, azimuth) and the distance between the geometric center and the centroid. However, for the 2 % reflective target, the distances between the geometric center and the centroid was clearly larger than the distances for the other reflectivities, see Figure 41. The probable reason for the larger distances is the irregular distribution of the points on the 2 % reflective target causing a shift in the centroid location.

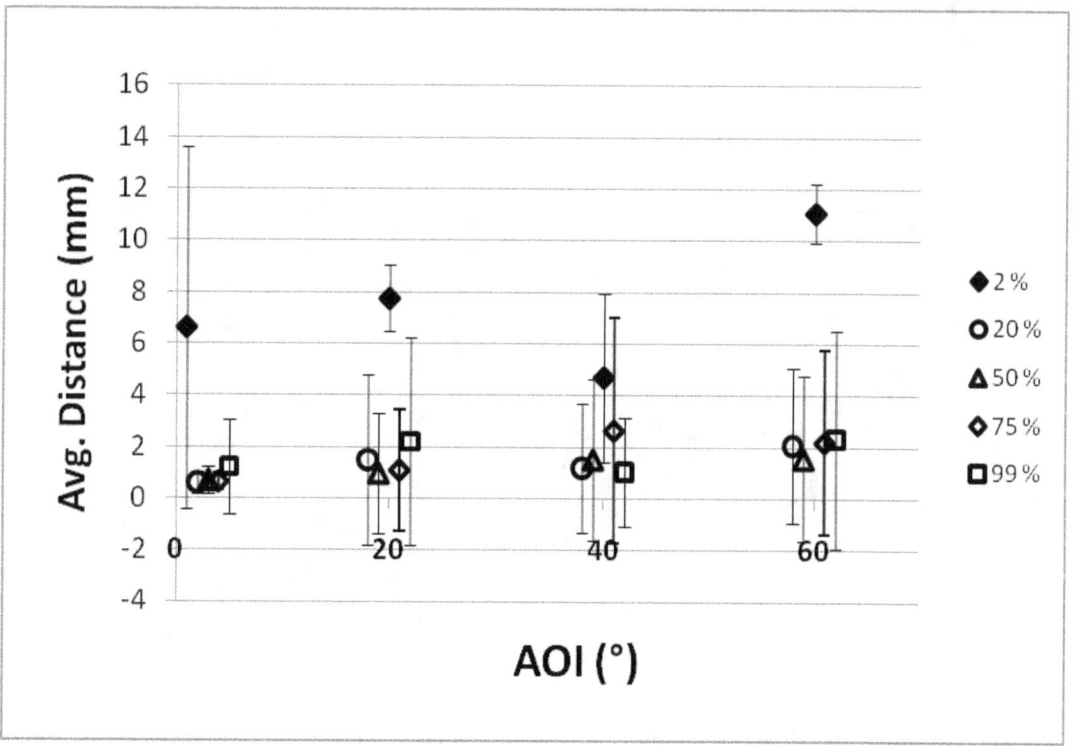

Figure 41. Average distance between geometric center and centroid vs. AOI. Points represent the averages at a given AOI and reflectivity over all ranges and azimuths. The error bars are the standard deviations of the means. Note: AOI values are offset from 0°, 20°, 40°, and 60° for clarity.

Some reasons why the centroid or geometric centers are incorrectly located include:

- The irregular patterns of the points on the target (see Figure 42) affects the location of the centroid. These patterns are caused by the instrument or the object surface as noted for the 2 % reflective target.
- The algorithm for locating the geometric center:
    - For scans where there was sparser data, it was noted that the automated algorithm included points on the target holder causing it to shift the geometric center up or down. Solutions include modifications to the algorithm to detect such conditions or to remove the points from the target holder before proceeding with the determination of the geometric center.
    - For scans when the target was rotated, the geometric center was shifted left or right. This was caused by the increased number of points on the edge of the target closer to the 3D imaging instrument. Again, modifications to the algorithm could be made to minimize this error.

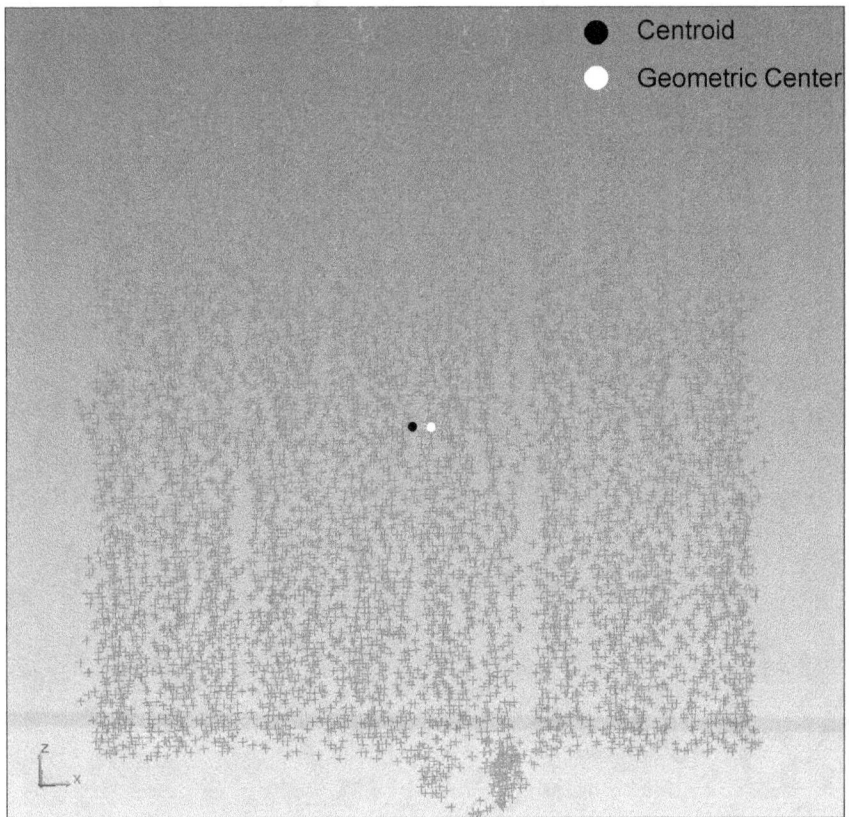

Figure 42. Irregular distribution of points on target. Note the bands where there are no data points.

5. Summary and Conclusions

An extensive set of experiments were conducted to evaluate the effects of various factors on the range error and noise of a 3D imaging instrument. The factors evaluated were: range, AOI, reflectivity, azimuth angle, method of obtaining range measurement (measure single point on target or scan the target), and target type (planar vs. spherical). A total of 190 tests were conducted.

The results show that the average errors (averaged over all AOIs, ranges, reflectivities, and azimuths) were within the manufacturer-specified range uncertainty. They also show that AOI and reflectivity have the greatest effect on range error. Reflectivity and range have the greatest effect on range noise (taken as the RMS of the residuals of the plane fit). However, the results from the 2 % reflective target skewed the results. If the data from the 2 % reflective target are omitted, then AOI has the largest effect on range error, and range has the largest effect on range noise. More specific findings include:

- Range error:
  - o Increases significantly for AOI = 60°
  - o For Method 1A, the instrument overestimated the range for ranges ≤ 60 m (or 0.3 R_{max}) and underestimated the range for ranges ≥ 110 m (or 0.55 R_{max}) with the larger errors occurring at ranges ≥ 110 m. For Method 1B, the instrument underestimated the range for all ranges.
  - o Larger range errors were obtained for the 2 % reflective target
  - o Azimuth has an insignificant effect
- Range noise
  - o Relatively stable for AOI $\leq 20°$ and increased for AOI of 40° and 60°
  - o Increases linearly with range
  - o 2 % reflective target significantly increases range noise
  - o Azimuth has no effect

For Method1B, the tests were reproducible for AOIs from 0° to 60° over all ranges and reflectivities. For Method 1A, the tests were reproducible for AOIs of 0° and 20° over all ranges and reflectivities. The combination of longer range and high AOI (60°) reduces the chances for reproducibility.

Based on the comparison of range errors, the significance of obtaining the range measurements, scanning target (Method 1A and 1B) and measuring single points on a target (Method 2) is strongly dependent on the AOI. Methods (1A and 1B) and Method 2 are not statistically different at the 5 % level for AOIs of 0° and 20°.

Of the two methods to define a point on a planar target, line-plane intersection (Method 1A) vs. geometric center (Method 1B), the geometric center method is recommended over the line-plane intersection method. The test results obtained using Method 1B were more reproducible than those using Method 1A. Also, the method for obtaining the geometric center is simpler, more straightforward, and yielded results that were more "scientifically rational", i.e, the smallest range error occurred at the range that is closest to the distance at which the laser beam was focused.

A comparison of the use of the centroid vs. the geometric center to define the point on the plane indicates that both methods would produce similar results for the majority of the cases. However, the use of the geometric center is preferred because the location of the centroid is more adversely affected by irregular point distribution (e.g., areas of missing points).

Statistically, the RMS of the plane fit and the standard deviation of single measurements are different at the 5 % level with the RMS values being greater than the standard deviation for the majority of the cases. On average, the RMS of the plane fit is 1.6 mm ($s = 0.2$ mm) greater than the standard deviation of single point measurements. The larger differences between the two representations of range noise occur at the longer ranges.

A planar target (single point measurements) and a spherical target would result in range errors that are considered statistically equivalent at the 5 % level based on the t test. A planar target (scanned) and a spherical target would result in range errors that are considered statistically different at the 5 % level based on the t test. However, on average, the range error for the planar target (scanned) is 1.7 mm greater than for the spherical target. A comparison of this value with the instrument uncertainty of 7 mm would indicate that the planar (scanned) and spherical targets may be considered equivalent.

Acknowledgements

The authors would like to thank Mr. Nicholas Philippon, a former guest researcher at NIST, and Mr. Jonathan Vega, a former summer student at NIST, for their help in conducting the experiments.

References

1. ASTM Standard E2544 2009a, *Standard Terminology for 3D Imaging Systems*, ASTM International: West Conshohocken, PA.
2. Pfeffer, C., *How good is your Scan? The state of 3-D Imaging Standards*, in *26th Annual Conference of the Coordinate Metrology Systems Conference (CMSC)*. 2010: San Antonio, TX.
3. Beraldin, J.-A., et al., *Traceable 3D Imaging Metrology*, in *Annual IS&T/SPIE on Electronic Imaging*. 2007, SPIE: San Jose, CA.
4. Cheok, G.S., *Proceedings of the LADAR Calibration Facility Workshop, June 12-13, 2003, NISTIR 7054*. 2003, National Institute of Standards and Technology: Gaithersburg, MD.
5. Hiremagalur, J., et al., *Creating Standards and Specifications for the Use of Laser Scanning in CALTRANS Projects*. 2007, University of California at Davis. p. 93.
6. Cheok, G.S., A.M. Lytle, and K.S. Saidi. *ASTM E-57 3D Imaging Systems Committee: An Update*. in *Defense and Security Symposium*. 2008. Orlando, FL: SPIE.

7. Cheok, G.S., S. Leigh, and A. Rukhin, *Calibration Experiments of a Laser Scanner*, in *NISTIR 6922*. 2002, National Institute of Standards and Technology. p. 117.
8. Boehler, W., M.B. Vicent, and A. Marbs. *Investigating Laser Scanner Accuracy*. in *XIXth CIPA Symposium*. 2003. Antalya, Turkey.
9. Reshetyuk, Y., *Investigation and calibration of pulsed time-of-flight terrestrial laser scanners*. 2006, Royal Institute of Technology: Stockholm. p. 152.
10. Beraldin, J.-A. and M. Gaiani, *Evaluating the Performance of Close Range 3D Active Vision Systems for Industrial Design Applications*, in *SPIE: Electronic Imaging*. 2005: San Jose, CA.
11. Salo, P., O. Jokinen, and A. Kukko, *On the calibration of the distance measuring component of a terrestrial laser scanner*, in *The International Archives of the Photogrammetry, Remote Sensing and Spatial Information Sciences*. 2008: Beijing. p. 1067-1071.
12. Voegtle, T., I. Schwab, and T. Landes, *Influences of different materials on the measurements of a terrestrial laser scanner (TLS)*, in *The International Archives of the Photogrammetry, Remote Sensing and Spatial Information Sciences*. 2008: Beijing. p. 1061-1066.
13. Lichti, D.D., *Error modelling, calibration and analysis of an AM-CW terrestrial laser scanner system*. ISPRS Journal of Photogrammetry and Remote Sensing, 2006. 61: p. 307-324.
14. Kersten, T., H. Sternberg, and E. Stiemer. *First Experiences with Terrestrial Laser Scanning for Indoor Cultural Heritage Applications Using Two Different Scanning Systems*. in *ISPRS working group V/5 Panoramic Photogrammetry Workshop*. 2005. Berlin, Germany.
15. Cheok, G.S., K.S. Saidi, and M. Franaszek. *Target Penetration of Laser-Based 3D Imaging Systems*. in *SPIE 21st Annual Symposium on Electronic Imaging - 3D Imaging Metrology*. 2009. San Jose, CA.
16. Press, W.H., et al., *Numerical Recipes in C*. 2nd ed. 1995: Cambridge University Press.

Appendix A: Experiment Test Settings and Results

	Test	Range X1 (m)	Reflectivity X2 (%)	AOI X3 (°)	Azimuth X4	RMS of fit		Error Method 1A		Error Method 1B	
						Avg. (mm)	Std dev (mm)	Avg. (mm)	Std dev (mm)	Avg. (mm)	Std dev (mm)
Day 1 (Test Set A)	1	15	99	0	1 ≈ 1°	1.6	0.04	3.2	0.1	3.3	0.2
(6/10/08)	2	15	99	40	1	1.3	0.02	0.1	0.2	-2.1	0.1
	3	15	75	20	1	1.2	0.01	1.9	0.1	2.4	1.8
	4	15	75	60	1	1.4	0.03	-2.2	0.3	-18.1	1.2
	5	15	50	40	1	1.3	0.02	1.8	0.0	-2.4	0.1
	6	15	50	0	1	1.2	0.01	1.8	0.1	1.7	0.1
	7	15	20	60	1	1.9	0.04	-2.0	0.1	-17.2	0.7
	8	15	20	20	1	1.5	0.05	3.0	0.1	2.7	0.2
	9	110	75	40	1	3.1	0.07	-2.4	0.3	0.5	0.5
	10	110	75	0	1	2.6	0.02	0.1	0.1	0.1	0.1
	11	110	50	60	1	5.1	0.10	-7.2	0.9	-7.0	3.6
	12	110	50	20	1	3.4	0.05	-1.5	0.2	-2.1	0.5
	13	110	20	0	1	4.9	0.15	-2.4	0.1	-2.4	0.1
	14	110	20	40	1	5.6	0.03	-5.8	0.6	-4.8	4.1
	15	110	99	20	1	2.4	0.05	2.1	0.2	2.7	1.3
	16	110	99	60	1	4.3	0.19	-2.3	0.6	2.5	1.6
Day 2 (Test Set A)	17	60	50	0	1	1.8	0.02	3.2	0.2	3.1	0.1
(6/11/08)	18	60	50	40	1	2.1	0.03	2.2	0.2	0.8	1.5
	19	60	20	20	1	2.7	0.02	1.2	0.2	0.5	0.3
	20	60	20	60	1	3.5	0.05	-0.3	0.3	-12.1	2.0
	21	60	99	40	1	1.8	0.01	3.0	0.3	2.8	0.1
	22	60	99	0	1	1.5	0.02	2.8	0.1	2.8	0.0
	23	60	75	60	1	2.4	0.07	2.5	0.3	-6.2	5.4
	24	60	75	20	1	1.5	0.02	2.0	0.2	-0.1	0.9
	25	160	20	40	1	8.8	0.16	-4.3	0.6	-5.3	1.2
	26	160	20	0	1	7.3	0.10	-3.1	0.1	-3.2	0.1
	27	160	99	60	1	6.9	0.21	-2.1	2.0	-3.2	3.3
	28	160	99	20	1	3.6	0.09	2.2	0.1	10.3	0.5
	29	160	75	0	1	3.9	0.05	-0.1	0.1	-0.2	0.3
	30	160	75	40	1	4.8	0.08	-0.3	0.5	-1.3	4.0
	31	160	50	20	1	5.1	0.04	-2.0	0.1	-3.1	2.0
	32	160	50	60	1	7.4	0.26	-8.3	0.6	-10.9	3.0
Day 3 (Test Set A)	33	110	20	20	1	5.2	0.16	-2.1	0.4	-2.2	2.8
(6/11/08)	34	110	20	60	1	6.8	0.50	-10.4	0.4	-8.6	5.2

	Test	Range X1 (m)	Reflectivity X2 (%)	AOI X3 (°)	Azimuth X4	RMS of fit		Error Method 1A		Error Method 1B	
						Avg. (mm)	Std dev (mm)	Avg. (mm)	Std dev (mm)	Avg. (mm)	Std dev (mm)
	35	110	50	40	1	3.8	0.02	-0.8	0.5	-3.5	1.0
	36	110	50	0	1	3.3	0.10	1.5	0.1	1.5	0.1
	37	110	75	60	1	4.5	0.16	-7.0	0.3	-7.9	2.9
	38	110	75	20	1	2.7	0.01	-0.4	0.1	0.4	1.3
	39	110	99	0	1	2.4	0.04	3.2	0.1	3.1	0.1
	40	110	99	40	1	3.0	0.02	-1.1	0.1	0.0	0.3
	41	15	50	60	1	1.5	0.02	-2.9	0.0	-19.2	2.4
	42	15	50	20	1	1.2	0.00	-0.1	0.0	-0.2	0.1
	43	15	75	0	1	1.3	0.07	-0.6	0.0	-0.8	0.1
	44	15	75	40	1	1.2	0.03	-0.7	0.0	-3.8	2.1
	45	15	99	20	1	1.4	0.01	0.6	0.1	0.7	0.5
	46	15	99	60	1	1.4	0.02	-3.9	0.2	-17.2	5.9
	47	15	20	40	1	1.7	0.03	0.0	0.2	-1.1	3.4
	48	15	20	0	1	1.4	0.01	1.9	0.0	1.7	0.0
Day 4 (Test Set A)	49	160	75	20	1	4.1	0.02	-1.1	0.2	0.5	2.3
(6/12/08)	50	160	75	60	1	7.0	0.18	-9.5	2.1	-7.9	7.8
	51	160	99	40	1	4.5	0.07	-0.4	0.4	-1.4	1.9
	52	160	99	0	1	3.3	0.04	3.4	0.1	3.4	0.1
	53	160	20	60	1	11.1	0.14	-9.1	1.3	-13.3	3.0
	54	160	20	20	1	7.7	0.09	-4.8	0.0	-5.2	0.2
	55	160	50	0	1	4.9	0.07	-1.1	0.3	-1.1	0.3
	56	160	50	40	1	5.8	0.19	-6.0	0.3	-5.7	4.1
	57	60	99	60	1	2.5	0.03	0.4	0.2	1.9	4.8
	58	60	99	20	1	1.5	0.03	2.9	0.2	0.2	0.6
	59	60	75	0	1	1.5	0.02	1.5	0.2	1.5	0.2
	60	60	75	40	1	1.8	0.02	1.1	0.3	0.5	4.1
	61	60	50	20	1	1.9	0.05	1.3	0.1	-0.4	1.2
	62	60	50	60	1	2.8	0.11	-1.0	0.4	-9.4	5.3
	63	60	20	40	1	2.9	0.03	0.1	0.2	-4.1	1.4
	64	60	20	0	1	2.6	0.07	0.9	0.2	0.9	0.2
Day 5 (Test Set A)	65	160	99	0	2 ≈ 240°	3.4	0.05	3.3	0.1	3.2	0.2
(6/12/08)	66	160	99	40	2	4.3	0.05	-0.4	0.5	-3.4	1.6
	67	160	75	20	2	4.1	0.07	-0.6	0.1	-0.4	1.4
	68	160	75	60	2	6.6	0.41	-5.9	2.0	-2.0	7.7
	69	160	50	40	2	5.6	0.07	-3.4	0.2	-2.2	1.1
	70	160	50	0	2	5.0	0.23	-1.3	0.2	-1.3	0.2

	Test	Range X1 (m)	Reflectivity X2 (%)	AOI X3 (°)	Azimuth X4	RMS of fit		Error Method 1A		Error Method 1B	
						Avg. (mm)	Std dev (mm)	Avg. (mm)	Std dev (mm)	Avg. (mm)	Std dev (mm)
	71	160	20	60	2	11.1	0.42	-1.1	1.8	-13.0	5.8
	72	160	20	20	2	7.5	0.06	-4.1	0.2	-3.7	1.8
	73	15	75	40	2	1.3	0.03	0.9	0.0	-0.9	1.6
	74	15	75	0	2	1.3	0.03	0.2	0.1	0.2	0.1
	75	15	50	60	2	1.5	0.05	-1.2	0.1	-23.7	0.9
	76	15	50	20	2	1.2	0.03	0.8	0.1	1.1	0.8
	77	15	20	0	2	1.4	0.03	2.4	0.1	2.4	0.3
	78	15	20	40	2	1.6	0.04	1.1	0.0	0.3	2.6
	79	15	99	20	2	1.4	0.01	2.3	0.2	1.8	1.1
	80	15	99	60	2	1.4	0.03	-0.5	0.1	-2.2	1.2
Day 6 (Test Set A)	81	110	50	0	2	3.2	0.05	0.8	0.1	0.7	0.1
(6/13/08)	82	110	50	40	2	3.9	0.03	-3.6	0.3	-0.7	0.6
	83	110	20	20	2	5.0	0.08	-2.5	0.3	-3.6	2.1
	84	110	20	60	2	6.6	0.16	-5.5	0.8	-13.7	2.0
	85	110	99	40	2	3.0	0.09	-3.5	2.0	0.3	6.7
	86	110	99	0	2	2.3	0.06	3.2	0.0	3.2	0.0
	87	110	75	60	2	4.5	0.03	-7.8	0.6	-9.5	3.4
	88	110	75	20	2	2.8	0.06	-1.2	0.1	0.5	0.2
	89	60	20	40	2	2.9	0.03	3.0	0.3	-3.6	2.5
	90	60	20	0	2	2.6	0.04	1.6	0.1	1.6	0.1
	91	60	99	60	2	2.4	0.06	-1.4	0.5	-3.4	3.5
	92	60	99	20	2	1.6	0.02	3.7	0.1	3.2	1.4
	93	60	75	0	2	1.5	0.01	1.6	0.1	1.5	0.1
	94	60	75	40	2	1.8	0.02	2.9	0.1	0.6	3.1
	95	60	50	20	2	1.9	0.02	0.1	0.0	1.2	1.0
	96	60	50	60	2	2.7	0.01	-2.2	0.5	-10.0	2.8
Day 7 (Test Set A)	97	15	20	20	2	1.5	0.03	3.5	0.1	3.4	0.4
(6/13/08)	98	15	20	60	2	2.0	0.09	0.3	0.1	-23.8	1.9
	99	15	50	40	2	1.3	0.01	0.8	0.2	2.3	0.5
	100	15	50	0	2	1.2	0.03	1.0	0.2	1.0	0.2
	101	15	75	60	2	1.4	0.04	-1.6	0.1	-10.3	8.2
	102	15	75	20	2	1.2	0.03	0.3	0.0	1.3	1.3
	103	15	99	0	2	1.6	0.04	1.3	0.1	1.2	0.0
	104	15	99	40	2	1.3	0.05	-0.4	0.1	-3.6	6.5
	105	160	50	60	2	7.9	0.26	-8.4	0.8	-7.2	5.0

	Test	Range X1 (m)	Reflectivity X2 (%)	AOI X3 (°)	Azimuth X4	RMS of fit		Error Method 1A		Error Method 1B	
						Avg. (mm)	Std dev (mm)	Avg. (mm)	Std dev (mm)	Avg. (mm)	Std dev (mm)
	106	160	50	20	2	5.2	0.12	-3.1	0.2	-2.6	2.0
	107	160	75	0	2	3.9	0.03	-0.3	0.1	-0.3	0.2
	108	160	75	40	2	4.9	0.05	-1.0	0.8	-2.6	3.3
	109	160	99	20	2	3.6	0.05	1.1	0.4	2.1	0.5
	110	160	99	60	2	6.8	0.41	-4.7	1.6	-9.7	4.2
	111	160	20	40	2	8.9	0.33	-4.4	0.3	-3.2	2.0
	112	160	20	0	2	7.4	0.13	-3.8	0.1	-3.7	0.1
Day 8 (Test Set A) (6/16/08)	113	60	75	20	2	1.5	0.02	1.7	0.1	2.2	1.1
	114	60	75	60	2	2.4	0.03	-0.4	0.3	-3.9	7.8
	115	60	99	40	2	1.8	0.03	1.2	0.1	1.2	2.0
	116	60	99	0	2	1.5	0.01	2.9	0.1	3.1	0.1
	117	60	20	60	2	3.4	0.15	-2.3	0.3	-10.5	9.3
	118	60	20	20	2	2.6	0.03	0.6	0.1	-0.7	1.7
	119	60	50	0	2	1.8	0.02	0.9	0.3	0.9	0.2
	120	60	50	40	2	2.1	0.08	-0.4	0.3	0.2	1.0
	121	110	99	60	2	4.2	0.07	-8.8	0.4	-3.1	8.9
	122	110	99	20	2	2.5	0.03	1.0	0.1	1.6	1.8
	123	110	75	0	2	2.6	0.02	-0.1	0.1	-0.2	0.1
	124	110	75	40	2	3.3	0.04	-3.7	0.3	-0.5	1.7
	125	110	50	20	2	3.4	0.02	-1.9	0.1	-1.9	0.4
	126	110	50	60	2	5.0	0.03	-7.7	0.9	-8.1	4.1
	127	110	20	40	2	5.7	0.09	-5.7	0.1	-6.6	1.6
	128	110	20	0	2	5.1	0.10	-2.7	0.0	-2.7	0.0
Test Set A: Repeats & 2 % target (6/16/08)	1r (103)[1]	15	99	0	2	1.6	0.01	3.2	0.1	3.2	0.1
	2r (97)	15	20	20	2	1.5	0.04	4.2	0.2	4.0	1.9
	3r (101)	15	75	60	2	1.4	0.02	1.6	0.2	-4.6	7.4
	4r (99)	15	50	40	2	1.4	0.06	0.8	0.1	-2.8	0.6
	5d[2]	15	2	0	2	6.3	0.09	-0.5	0.1	-0.6	0.1
	6r (92)	60	99	20	2	1.6	0.01	1.9	0.2	2.4	0.9
	7r (93)	60	75	0	2	1.5	0.02	0.5	0.0	0.6	0.1
	8r (89)	60	20	40	2	2.9	0.08	-1.7	0.1	-6.1	0.5
	9r (96)	60	50	60	2	2.8	0.07	-2.2	0.3	18.2	5.9
	10d	60	2	0	2	14.1	0.67	2.1	0.5	1.9	0.5
	11d	60	2	20	2	15.4	0.18	3.7	1.0	-0.4	1.1
	12d	60	2	40	2	17.1	0.84	5.6	1.6	-2.9	1.2
	13d	60	2	60	2	17.6	2.12	6.4	4.9	-20.1	2.2
	14r (81)	110	50	0	2	3.3	0.07	-0.7	0.2	-0.9	0.2

	Test	Range X1 (m)	Reflectivity X2 (%)	AOI X3 (°)	Azimuth X4	RMS of fit Avg. (mm)	RMS of fit Std dev (mm)	Error Method 1A Avg. (mm)	Error Method 1A Std dev (mm)	Error Method 1B Avg. (mm)	Error Method 1B Std dev (mm)
	15r (88)	110	75	20	2	2.8	0.03	-2.5	0.1	-0.9	0.6
	16r (85)	110	99	40	2	3.0	0.09	-5.0	0.1	-0.8	0.7
	17r (84)	110	20	60	2	6.5	0.16	-17.7	0.8	3.4	4.7
	18d	110	2	0	2	No data	No data	No data	No data	No data	No data
	19d	110	2	20	2	No data	No data	No data	No data	No data	No data
	20d*	110	2	40	2	Test not performed since no data was obtained for Test 19d.					
	21d*	110	2	60	2	Test not performed since no data was obtained for Test 19d.					
	22r (110)	160	99	60	2	6.4	0.20	-26.0	1.2	-10.9	2.3
	23r (108)	160	75	40	2	5.0	0.12	-13.2	0.3	-5.0	1.2
	24r (106)	160	50	20	2	5.1	0.08	-7.1	0.6	-2.5	2.3
	25r (112)	160	20	0	2	7.3	0.08	-4.6	0.1	-4.6	0.1
	26d*	160	2	0	2	No data	No data	No data	No data	No data	No data
Test Set B: Single point Measurements (6/17/08)		15	20	0	2	NA[3]	NA	2.0	1.2		
		15	20	20	2	NA	NA	2.1	1.1		
		15	20	40	2	NA	NA	3.4	0.9		
		15	20	60	2	NA	NA	4.3	2.0		
		15	99	0	2	NA	NA	4.5	1.2		
		15	99	20	2	NA	NA	2.7	1.6		
		15	99	40	2	NA	NA	2.5	1.5		
		15	99	60	2	NA	NA	2.6	1.6		
		60	20	0	2	NA	NA	-0.8	0.8		
		60	20	20	2	NA	NA	1.0	1.7		
		60	20	40	2	NA	NA	3.4	1.6		
		60	20	60	2	NA	NA	8.3	3.1		
		60	99	0	2	NA	NA	1.5	1.1		
		60	99	20	2	NA	NA	3.5	1.0		
		60	99	40	2	NA	NA	6.3	2.0		
		60	99	60	2	NA	NA	7.0	1.5		
		110	20	0	2	NA	NA	-3.8	1.4		
		110	20	20	2	NA	NA	-1.0	1.6		
		110	20	40	2	NA	NA	0.2	2.0		
		110	20	60	2	NA	NA	2.7	4.2		
		110	99	0	2	NA	NA	2.3	1.2		
		110	99	20	2	NA	NA	4.7	1.5		
		110	99	40	2	NA	NA	4.9	1.7		
		110	99	60	2	NA	NA	6.7	3.1		

Test	Range X1 (m)	Reflectivity X2 (%)	AOI X3 (°)	Azimuth X4	RMS of fit		Error Method 1A		Error Method 1B	
					Avg. (mm)	Std dev (mm)	Avg. (mm)	Std dev (mm)	Avg. (mm)	Std dev (mm)
	160	20	0	2	NA	NA	-4.4	0.9		
	160	20	20	2	NA	NA	-4.9	1.3		
	160	20	40	2	NA	NA	-5.5	1.7		
	160	20	60	2	NA	NA	-9.3	3.1		
	160	99	0	2	NA	NA	-0.1	0.9		
	160	99	20	2	NA	NA	3.2	1.7		
	160	99	40	2	NA	NA	5.4	1.5		
	160	99	60	2	NA	NA	6.2	3.5		
Test Set C: Spherical target (6/17/08)	15	51	NA	2	1.4	0.09	-0.2	0.0		
	33	51	NA	2	1.3	0.03	-0.7	0.0		
	60	51	NA	2	1.8	0.07	0.0	0.2		
	110	51	NA	2	3.6	0.08	-1.9	0.4		
	142	51	NA	2	4.7	0.08	-3.5	0.4		
	160	51	NA	2	5.3	0.42	-4.3	0.3		

Notes:
1. The suffix "r" means this is a repeat test. The number in parentheses is the test # that is repeated.
2. The suffix "d" means the target had a 2 % reflectivity.
3. NA = Not applicable as there was no fitting performed.

www.ingramcontent.com/pod-product-compliance
Lightning Source LLC
Chambersburg PA
CBHW081840170526
45167CB00007B/2857